SUSTAINABLE DEVELOPMENT AND THE ENERGY INDUSTRIES

Implementation and Impacts of Environmental Legislation

EDITED BY NICOLA STEEN

THE ROYAL INSTITUTE OF
INTERNATIONAL AFFAIRS
Energy and Environmental Programme

EARTHSCAN
Earthscan Publications Ltd, London

First published in Great Britain in 1994 by
Earthscan Publications Ltd, 120 Pentonville Road, London N1 9JN and
Royal Institute of International Affairs, 10 St James's Square, London SW1Y 4LE

Distributed in North America by
The Brookings Institution, 1775 Massachusetts Avenue NW,
Washington DC 20036-2188

A catalogue record for this book is available from the British Library.

ISBN 1 85383 210 3

Earthscan Publications Limited is an editorially independent subsidiary of Kogan Page
Limited and publishes in association with the International Institute of Environment and
Development and the World Wide Fund for Nature.

Printed in England by Clays Ltd, St Ives plc
Cover by Elaine Marriott

Research by the Energy and Environmental Programme is supported by generous contributions of finance and professional advice from the following organizations:

AEA Technology · Amerada Hess · Arthur D Little
Ashland Oil · British Coal · British Nuclear Fuels
British Petroleum · European Commission
Department of Trade and Industry · Eastern Electricity
Enterprise Oil · ENRON Europe · Exxon · LASMO
Mobil · National Grid · National Power · Nuclear Electric
Overseas Development Administration · PowerGen
Saudi Aramco · Shell · Statoil · Texaco · Total
Tokyo Electric Power Company

SUSTAINABLE DEVELOPMENT AND THE ENERGY INDUSTRIES

Proceedings of a workshop organized by the
Energy and Environmental Programme of the
Royal Institute of International Affairs
in collaboration with the
Fondazione ENI Enrico Mattei
held at Chatham House, London, on 19 November 1993,
with the generous support of
Sainsbury Monument Trust and
Green College Centre for Environmental
Policy and Understanding.

Contents

Foreword

Energy industries supply goods and services that make much of modern life possible, but often face strong environmental criticism. The energy industries have incurred high costs as a result of past environmental legislation, and some energy companies have vigorously opposed and often helped to block new legislation. The nexus between environmental policy, sustainability and the energy industries is fraught with difficulty and contention, but better understanding of the positions, interests and possibilities is essential if good policy is to emerge.

In an effort to learn from international experience and to further the dialogue and mutual understanding between perspectives that often seem irreconcilably opposed, on 19 November 1993 the Royal Institute of International Affairs held an international workshop on 'Sustainable Development and the Energy Industries'. Covering such a broad topic in a single day led to a compact programme but enabled high-level participation from many different perspectives: fossil fuel, electricity and chemical industries; environmental lobby groups; academics from universities and research institutes; and some government departments. To maximize the time available for discussion, papers were prepared and circulated in advance, with the main points presented to the workshop by the authors as a basis for discussion.

This edited volume presents the workshop papers together with an overview of the workshop themes and discussions by the editor. An Appendix contains an edited version of my welcoming speech to the workshop, which sets out more of the background. In preparing their papers for publication, the authors were encouraged to make revisions in the light of comments at the workshop, and any other remarks individual participants wished to relay. The papers have been edited for clarity, but not for accuracy; as workshop papers they have not been subject to the usual review process accorded to research reports by the Royal Institute of International Affairs. The views expressed are those of the individual authors and workshop participants, not of the RIIA or of the editor.

Yet the authors have approached the task with care and dedication to produce an unprecedented collection which not only presents the key issues but also illustrates the wide range of views with which policy will have to contend. Nilsson concludes that much tougher environmental standards, on a range of emissions, are still needed to meet scientific goals; Malin points out that big improvements have already been achieved, criticizes past excesses and questions whether the benefits of taking policy further would really justify the potentially rapidly rising control costs. Mitchell doubts the practicality, fairness and efficacy of relying on 'polluter pays' taxes before alternative technologies are readily available, whilst Cable views them as the most efficient and least trade-distorting approach, and von Weizsäcker advocates broader tax reform as the main driver of the technical innovation needed to meet long-term goals. Even interpretations of past experience differ significantly: contrast Sioshansi's caution about the pioneering 'Californian experience' with Weinberg's optimism. The proceedings thus provide not only a valuable collection of information and argument, but also an insight into the beliefs and perceptions of a range of leading actors involved internationally in the energy-environment debate.

The Energy and Environmental Programme is deeply indebted to the sponsors whose support made this work possible. The Monument Trust and Green College Oxford donated money to cover the costs of running the workshop, and we are most fortunate also to have been supported intellectually and financially by the Fondazione ENI Enrico Mattei, whose support enabled production of proceedings far beyond a normal workshop report. Most of all I would like to express gratitude to Nicola Steen, who organized the workshop from its inception and has edited the proceedings with care and dedication. Nicola made the rare transition from an administrative position on the Programme to that of Junior Research Fellow, and the workshop and proceedings demonstrate her considerable abilities in both roles.

April 1994   Dr Michael Grubb
                                   Head, Energy and Environmental Programme

Editor's Acknowledgements

As editor and organizer of the workshop, I would like to give my thanks to all participants of the workshop for taking part in the discussion, and special thanks to those who presented papers. These contributors provided their amended papers for the book, graciously fulfilling editing requests and meeting deadlines. Thank you, to the readers John Wybrew, Alessandro Lanza and David Wallace, for constructively suggesting improvements to the overview chapter; to Valerie Fogleman for advising on legal points; and to Walt Patterson for a crash course in writing. Thank you, Nicole Dando, for organizing and running the workshop with me and for efficiently and consistently managing the production of the book. Matthew Tickle and Margaret May have kept a calm watch over proceedings; Eileen Power copy-edited the book and Hannah Doe typeset it: I appreciate all of their work. Thanks to Silvan Robinson for initiating the workshop, and many thanks to Michael Grubb for the opportunity of editing the publication and for encouraging and directing me, with written additions, suggestions and support.

April 1994   Nicola Steen

Notes on Contributors

Sir Crispin Tickell, GCMG, KCVO

Warden of Green College, Oxford; Convener of the British Government Panel on Sustainable Development; Chairman of the International Institute for Environment and Development; and Chairman of the Climate Institute of Washington DC. Sir Crispin Tickell also holds other posts.

Sir Crispin was a member of the Diplomatic Service. He was Chef de Cabinet to the President of the European Commission (1977–80), Ambassador to Mexico (1981–83), Permanent Secretary of the Overseas Development Administration (1984–87), and British Permanent Representative to the United Nations (1987–90). He was also President of the Royal Geographical Society (1990–93).

Marcello Colitti

Since June 1993 he has been the Chairman of EniChem SpA, the Chemical Company of ENI Group. Previously he was Energy Advisor to the Chairman of ENI and Chairman of Ecofuel SpA, a Company of ENI Group; Chairman of Enichem Polimeri, a chemical and plastic Company of ENI Group; and Vice Chairman of AGIP SpA.

He is a member of the Oxford Energy Policy Group, and has published extensively and lectured on oil and economics. He has served as Chairman of the Industrial Advisory Board to the International Energy Agency, Paris (1982–3). He has a degree in law from the University of Parma.

Dr Lars J. Nilsson

Lars Nilsson is a Research Scientist at the Department of Environmental and Energy Systems Studies at Lund University and is currently a Visiting Research

Fellow at the Center for Energy and Environmental Studies at Princeton University. He holds an MSc in Engineering Physics and a PhD in Environmental and Energy Systems Studies from the Lund Institute of Technology, Lund University.

Dr Nilsson has worked primarily in the field of end-use oriented energy analysis and with assessments of energy efficient end-use technologies. He has also studied long-term trends in national energy intensities, nuclear proliferation issues, and regulatory and institutional aspects of the integration of intermittent sources of electricity.

Professor Thomas B. Johansson

Professor Thomas B. Johansson is Professor of Energy Systems Analysis at the University of Lund. He has been with the Department of Environmental and Energy Systems Studies at the University since 1985. He has a master's degree in engineering physics and his doctorate was in nuclear physics. He has written extensively in his field, including as co-editor of the UN-sponsored study *Renewable Energy: Sources for Fuels and Electricity*. Professor Johnasson was Chairman of the UN Solar Energy Group for Environment and Development, is a Convening Lead Author for the Energy Supply Mitigation Options for the IPCC and holds a number of similar positions.

Valerie M. Fogleman

Valerie Fogleman is a US attorney with the Reinsurance Division of the London solicitors, Barlow Lyde and Gilbert, and a member of the firm's Environmental Liability Group. Her responsibilities include advising on environmental law and pollution insurance matters. She is a member of the Texas State Bar and the American Bar Association.

Before joining Barlow Lyde and Gilbert in June 1992, she practised environmental law in Corpus Christi, Texas. She has a Juris Doctor degree, a Master of Science degree, and a bachelor's degree from Texas Tech University and a Master of Laws degree from the University of Illinois. She has written widely; her second environmental law book, *Hazardous Waste Cleanup, Liability and Litigation: A Comprehensive Guide to Superfund Law*, was published in November 1992.

Dr Vincent Cable

Head of the International Economic Programme, Royal Institute of International Affairs. Until September 1993 he was Acting Head of Socio-Economic Studies in Group Planning, Shell International, working on Shell's long-term scenarios.

Before joining Shell he had been an economist for businesses, in government, in international organizations and in universities. His primary interest is in international trade issues and in environmental economics (the latter with the Brundtland Commission). He has extensive experience of Asia (especially India), East Africa and Latin America.

John V. Mitchell

Chairman of the Energy and Environmental Programme at the Royal Institute of International Affairs. He is also Research Adviser to the Oxford Institute for Energy Studies.

John Mitchell retired in October 1993 from the position of Special Adviser to the Managing Directors of British Petroleum at its headquarters in London. From 1983 to 1990 he was Regional Coordinator for BP's affairs in the Western Hemisphere, managing links between BP headquarters, the Standard Oil Company, and other BP subsidiaries. He took part in the formation of BP America following the acquisition by BP of the minority shareholding in Standard Oil in 1987.

From 1978 to 1983 he was head of the Policy Review Unit, which was part of the staff of the Managing Directors of BP. He was also the company's senior economist and took part in a number of international collaborative studies of energy markets and taxation.

Clement B. Malin

Vice President, International Relations in the Corporate Communications Division of Texaco Inc. Malin joined Texaco in 1978 as Assistant to the Vice President – Strategic Planning and was appointed Director of the department later that year. He held various appointments Economics, including Vice President of International Relations for Texaco Europe, before assuming his present position in 1992.

Prior to joining Texaco, Malin served as Assistant Administrator for International Energy Affairs in the Federal Energy Administration from 1974 to 1977.

His previous oil industry experience, with the Mobil Corporation, included planning assignments in Europe, Latin America, the Middle East and Africa.

Malin is a member of the Council on Foreign Relations. He served as Chairman of the Industry Advisory Board to the International Energy Agency from 1984 to 1986. He holds a bachelor's degree in international relations from Dartmouth College and a master's degree in public affairs from Princeton University's Woodrow Wilson School of Public and International Affairs.

Anthony Baker

Anthony Baker is Head of Economics, British Coal. With degrees in mathematics and statistics, he joined the operational research team in the then UK National Coal Board in the early 1960s, where he undertook economic modelling and consulting work.

In the mid-1970s he set up and ran the Economic Assessment Service of IEA Coal Research, a collaborative project of the International Energy Agency. He later directed the whole IEA Coal Research programme, producing analysis with a wide perspective on the provision and economic use of coal internationally.

He was invited back to British Coal in 1987 to help manage the application of information technology to the reshaping of British Coal administration; in October 1991 he became Head of Economics.

Dr John C. Whitehead

Dr John C. Whitehead has been Head of the Coal Research Establishment since 1992. He graduated in chemical engineering, wrote a doctorate on coal liquefaction using supercritical gases and joined CRE in 1967.

Dr Fereidoon P. Sioshansi

Senior Project Manager in the Customer Systems Division of the Electric Power Research Institute (EPRI) in Palo Alto, California. He manages research projects on environmental aspects of demand-side management programmes and innovative rate design. He joined EPRI in 1991.

Dr Sioshansi worked as a Senior Research Scientist with Southern California Edison Company (SCE) in Rosemead, California for eleven years, where he served as Assistant to Manager of Steam Generation Division and Manager of R&D. He worked in a number of divisions including Advanced Engineering, R&D, System Planning, Power Contracts, and Power Supply. His responsibilities included forecasting, scenario analysis, strategic planning, R&D budgeting, and economic analysis. He served on several EPRI Project Advisory Groups and was Editor of SCE's Research Newsletter.

Dr Sioshansi is active in the International Association for Energy Economics and served as IAEE's Vice President for Publications and Editor for four years. He has participated in many professional associations and has published extensively. He has a BS and an MS degree in Civil and Structural Engineering from Purdue University and a PhD degree in Economics from the Krannert Graduate School of Management, Purdue University.

Françoise Garcia

An economist from the University of Paris, Panthéon Sorbonne (DEA, Economie de la Recherche-Développement), Françoise Garcia has been head of the Economic Division of the French Agency for Environment and Energy Management since 1992.

From 1983 to 1992 she worked for the French Agency for Energy Management, Economic Division, where she had particular responsibility for energy demand analysis and planning. From 1989 to 1990 she worked in the Office for Energy Efficiency of the Ministry of Energy and Resources of Quebec, Canada.

Dr Ernst Ulrich von Weizsäcker

Since April 1991 President of the Wuppertal Institute for Climate, Environment and Energy. Director of the Institute for European Environmental Policy in Bonn (IEUP) since 1984. Member of the Club of Rome.

Dr Weizsäcker studied and worked at various universities from 1969–1980. He has a first degree in Chemistry and Physics and a doctorate in Biology from Freiburg University. From 1975 to 1980 he was the Founding President of the new University of Kassel.

He was Director of the United Nations Centre for Science and Technology for Development, 1981–4, and from 1984 to 1991 he was Director of the Institute for European Environmental Policy in Bonn. In 1989 he received, together with the Norwegian Prime Minister Gro Harlem Brundtland, the Italian Premio de Natura Prize.

Professor von Weizsäcker has published numerous papers and books, including *Earth Politics*, 1994, and *Ecological Tax Reform* (co-authored with Jochen Jesinghaus), 1992.

Carl J. Weinberg

Recently retired from the Pacific Gas and Electric Company where he developed and directed an internationally respected energy research and development programme. He joined Pacific Gas and Electric Company in 1974 and had previously been a Colonel with the United States Air Force (1953–74), where his assignments involved bioscience management, and research and development in the United States, the Far East, and Europe.

Carl Weinberg is an internationally recognized spokesperson on energy efficiency and renewable energy issues, with a comprehensive view encompassing and integrating technical, regulatory, policy, and environmental perspectives. He serves on boards and working level committees of numerous energy efficiency and renewable energy organizations in the public and private sector. He received the Lifetime Achievement Award from the American Wind Energy Association in 1992 for advancing the cause of wind energy among the utilities nationwide.

He has master's degrees in Physics and Sanitary Engineering and a BSc in Civil Engineering. He has written widely, most recently a chapter on utility strategies in the UN-sponsored study *Renewable Energy: Sources for Fuels and Electricity*.

Katsuo Seiki

Katsuo Seiki is Executive Director, Global Industrial and Social Progress Research Institute (GISPRI). After graduating from the Faculty of Law, University of Tokyo in 1965 he joined the Ministry of International Trade and Industry (MITI). He was Deputy Director-General, Global Environmental Affairs, Minister's Secretariat when he left MITI in 1992 to take up his position at GISPRI.

Yutaka Tsuchida

Yutaka Tsuchida has been Manager at GISPRI since June 1993 when he was seconded from NKK Corporation. He has a degree from the Faculty of Engineering, Hokkaido University.

Dr Michael Grubb

Dr Michael Grubb is Head of the Energy and Environmental Programme at the Royal Institute of International Affairs, where he is responsible for directing a wide range of research on international energy and environmental issues. He is well known for his work on the policy implications of climate change, with publications including a report on the negotiating issues and options, a two-volume international study entitled *Energy Policies and the Greenhouse Effect* and a range of journal publications on economic and political aspects of the problem. He has also led a collaborative study on *Emerging Energy Technologies*, and most recently published a book examining the outcome and implications of the Rio 'Earth Summit' UN Conference on Environment and Development.

In addition, he is an adviser to a number of international organizations and studies, particularly concerning economic and policy aspects of climate change.

Dr Grubb gained a degree in Natural Sciences from Cambridge University, and subsequently worked at the Cavendish Laboratory in Cambridge and at Imperial College in London, where his studies resulted in various publications on the planning of electricity systems and the economic prospects for renewable energy sources.

Nicola Steen, Editor
Junior Research Fellow, Energy and Environmental Programme, Royal Institute of International Affairs. She has worked with EEP in various positions since 1989, being the Programme Administrator 1990–92. Previously she worked in Sudan for two years and from May 1994 she will be Policy Analyst at the Association of Independent Electricity Producers, London.

She holds an MSc in Economics from Birkbeck College, University of London, and a degree in German and French from University of Reading.

Sustainable Development and the Energy Industries: Implementation and Impacts of Environmental Legislation

Overview of Workshop Discussions

Nicola Steen

Introduction

The RIIA workshop on Sustainable Development and the Energy Industries brought together a wide range of perspectives on energy and environmental issues. The background to the workshop has been set out in the Foreword and Appendix. Subsequent chapters contain the revised and edited workshop papers. This overview aims to summarize and organize some of the important themes raised by the presentations at the workshop, and others brought out in the discussion. Discussion was inevitably limited by the time available, but nevertheless covered a wide area.

Since the 1970s concerns about the environment have led to considerable legislation being put in place, much of which affects the energy industries. Some participants in the workshop wanted more. Some were critical of the amount in place. Many criticized the methods or results of the legislation's implementation, and wanted future laws to learn from the problems caused in the past. But everyone assumed more legislation would be passed.

Some of the economic and structural consequences of earlier legislation do not seem to have been sufficiently analysed before the legislation was implemented. Fogleman[1] gave several simple examples of insufficient foresight: in February 1991, 13 of the top 20 property and casualty insurers reported

---

[1] Names in the text (sometimes in parentheses) refer readers to papers given at the workshop by the named participants and included in this publication in edited form.

Comments made in discussion at the workshop are not attributed, as the meeting was held under the Chatham House Rule of confidentiality (amended) to encourage free and frank discussion.

having approximately 50,000 claims and 2,000 lawsuits pending over pollution claims; one estimate of the *administrative* cost of complying with just one aspect of the 1990 Amendments to the US Clean Air Act is $500 million per year; a further study has estimated that the US oil industry must spend over $166 billion between 1991 and 2010 to comply with existing and anticipated environmental requirements. These figures do not take account of the risk that a firm's reputation will be damaged if it is found to be the cause of pollution. Legislation has proved to be difficult to enforce and complicated and costly to administer. The Rand Institute of Civil Justice has undertaken a study of five of the Fortune 500 companies. It reported that 27 per cent of the money assigned to clean-up by companies was used for transaction costs, mainly legal fees. In addition, 88 per cent of the money the insurers have paid because of actions related to environmental problems has been for legal fees. Not surprisingly, industries are nervous about the prospect of further regulation.

This legislation has, however, been demanded. It has been adopted because of public concern for the environment and realization of a need for a new type of development: sustainable development. Sir Crispin Tickell, in his opening speech, pointed to the need to change vocabulary and definitions. He criticized the tendency to view growth as necessarily good, and to see the only alternative to growth along existing trends as going backwards. Instead of using the term development as a synonym for growth, Sir Crispin suggested thinking of development as a 'change of potential'; a change for the better.[2]

Discussion centred on how environmental legislation would impinge and had impinged on the energy industries within a context of industries and government aiming for a sustainable energy policy. Responding to past legislation and anticipating further laws, industries have already started to restructure, and this process involves making changes to the manner and the direction of their growth. Examples of this were presented at the workshop, with suggestions as to how and where further changes might take place. Discussion recognized that companies do not necessarily have to wait to act until legislation is in place – and neither do countries have to wait for

---

[2] Although some would say economic literature has distinguished between 'growth' and 'development' since the 1950s, the terms have been and still often are interchanged in general usage.

international agreements – in order to respond. There are win-win situations, in which an action can have both economic and environmental benefits: in these cases action that pre-empts legislation will pay dividends.

This overview looks at what has changed and what is changing, and suggests areas where change will be necessary if energy industries are to meet goals of sustainable development.

What has changed?

Legislation has limited and in some cases reduced pollution: for example, in his paper Malin points out that the exhaust emissions of several pollutants from a new car in 1993 in the US were up to 96 per cent less than in 1970. Regulations have eliminated use of leaded petrol by new cars, and drastically reduced sulphur dioxide emissions from new power stations. Set against these improvements has been the impact of continued growth and ageing stock, so that the problems of air quality and sulphur emissions are still not fully resolved. (Nilsson)

However, environmental legislation has tended to have unanticipated results. Retrospective legislation has been passed in the US and has proved to be extremely problematic. Many countries have adopted the 'polluter pays' principle: this would suggest that where action is taken to clean up pollution the companies which created the problem should pay for the remedial action. However, in the US it is often the insurers who have had to pay. They have paid billions of dollars because of this legislation, put in place in the 1980s but covering actions before that time. The firms that insured industries had no idea of the risks they were taking when policies were sold in the 1960s and earlier. (Fogleman)

Besides the problem that final costs have been larger than anticipated, determining who is responsible for environmental damage can be difficult. US legislation has allowed for joint and several liability, which has often meant immense lawyers' bills as one company searches for and sues another.

As legislation moves more towards strict liability, according to which companies can be held responsible for environmental damage even if they were using the best available technologies at the time the damage occurred, acceptable categories of defence must be established. (Mitchell) Without such categories companies face potentially unlimited liability and enormous, un-anticipated bills.

Changes under-way

Many issues were discussed and nearly all were viewed as subject to change: either of necessity, or as a result of pressures. When change was considered necessary it was because it was seen as being either desirable for the common good or part of the most efficient economic path. Where pressures were considered likely to precipitate change, there was discussion as to whether the changes were indeed necessary; what would be the results of taking and not taking action; and, of course, investigation of the source of the pressures.

Technologies, accounting methods and indicators are changing, as are larger industrial and international structures.

Technologies

The range of technologies available and employed in the energy industries has changed over the last 20 years: gas turbines are replacing boilers and steam turbines; gas is replacing coal; renewable energy sources are maturing. In addition to tapping new resources, technologies have been developed to use the same resources more efficiently. Coal is a prime example: pressurized fluidized bed combustion and integrated gasification combined cycle plant have higher conversion efficiencies than traditional technologies, and dramatically reduce the pollution. (Baker and Whitehead) For such progress in cleaner and more energy-efficient technologies to continue, von Weizsäcker refers to the challenge of concentrating technological progress on productivity of energy rather than of labour.

Technological advances in general mean that companies are closer to their customers, leading to an emphasis on increased quality of services rather than selling increased quantities of products.

The potential impact of new technologies is great: it is likely that the technologies developed in industrialized countries will be exported and adopted in the developing world.

Methods of appraisal and investment planning in electricity[3]

Traditional accounting methods used in energy industries – especially utilities – have been developed to work within and for a large, centralized, often monopolistic or oligopolistic industry. Even on the supply side, many of the new technologies are small, modular units rather than large, centralized power generation systems which have been the norm. These smaller units alleviate the problems of lumpy investment, where the only option to increase supply is to add a large, expensive unit to the energy system.

New appraisal methods need to reflect that whereas old technologies often had economies of size, modular technologies tend to have economies of manufactured production. The full economies of production are unlikely to be achieved until the technologies' potentials are more fully exploited. In addition, the long timescales of large plant investment can mean capital being tied for long periods. This can create a self-fulfilling prophecy, that other technologies are not a viable investment option. (Weinberg)

A further example of change is that conventionally the process of transmission and distribution has been considered an overhead, a fixed part of an investment plan. This is being challenged by new transmission and interfacing technologies and smaller dispersed electricity generation units.

Implementing projects with the highest short-term paybacks may not be the same as – indeed, may not be – a strategy that has the highest long-term payback. By definition, a long-term perspective is necessary if we are working within a framework of sustainable development. In addition to taking assured 'win–win' actions, by taking a long-term approach and avoiding excessive impacts on the environment, firms could anticipate some of the areas that future environmental legislation might target. They could reduce potential liabilities, and also give themselves a possible future market advantage over their competitors.

Industrial structures and roles

Industrial restructuring can have many sources. Governments may seek to inject competition, for example through legislation for third-party access to

---

[3] It is surprising that the topic of 'sustainable development indicators' was not mentioned at the workshop, perhaps because their recent development has focused more upon national- than industry-level measures.

gas and electricity networks; they may intervene in markets to encourage demand-side management (DSM). Companies have to respond to regulation and government direction, but they also initiate restructuring. Objectives might include greater organizational efficiency through decentralization; or attracting new business in expanding markets, often in industrializing countries.

As resource owning and selling companies realize that their customers want low cost, high quality energy services, it is likely that these industries will see structural changes: activities will alter, and the large corporate model will break down into smaller, more dispersed activities.[4] (Weinberg, Figure 9.) This has already happened in the gas industry in the US and Sweden, where third-party access has led companies to open up their trade – previously often monopolistic – to competition. The UK, Italy, India and Brazil are opening their utilities to private investment. Oil companies are already largely competitive and often highly taxed, but they too are changing activities as emphasis moves to providing services.

Energy industries will have a role as providers of information: when customers are involved in the equation, information is important. If people are informed about an action they are more likely to carry it out – especially if they learn that the activity can bring rewards. Utilities devise their efficiency schemes so that it is in their interest that as many customers as possible adopt the schemes.

The cost of implementing such DSM projects includes advertising costs and – if demand is successfully reduced and rates not increased – potential loss of revenue. When the Californian regulators insisted that least cost methods be used – comparing plant expansion with DSM – Californian utilities pointed out that DSM would lower their profits if rates were unaltered. They bargained for the regulatory regime to be changed to allow recovery of their costs, and *also* successfully bargained for incentives and rewards for initiating the changes. (Sioshansi) It is this changed regulatory regime that has led Californian utilities to take much of the responsibility for disseminating new information and for undertaking energy efficiency schemes.

There were comments that those companies which would gain most from the current situation of external demands for sound environmental management

---

[4] Energy services – comfort, illumination, cooked food, mobility and so on – are what people actually want, not coal, oil, gas or electricity.

would be those which were willing and able to change course with the greatest speed and effect. One participant commented that even if change was seen to be needed within large industries, the organizations were like tankers that were difficult to steer out of the course upon which they had embarked.

A further problem was touched upon: any body in control of a centralized energy system would probably lose political power and influence if modular, more diverse structures and independent power production were introduced into the system. Such a body would initially have to sanction the installation of new modular systems. This might be a particular problem in industrializing countries where control over the energy supply is often an important tool for the government.

International structures
Investment, markets and economic growth Colitti argues that consumerism in industrialized countries is now reaching saturation point, so although technology continues to advance, demand and growth in many of the rich countries is sluggish, resulting in economic stagnation. To move out of stagnation these economies have to expand. It was argued that economies are too rigid for Keynesian economics to be effective, and in any case governments feel constricted by deficits and debt. Expansion will happen where growth is demanded – namely in industrializing countries. Increased direct investment in these poorer countries would allow for development there and could stimulate economies in the richer countries.[5]

Both Colitti and von Weizsäcker argue, for example, that current financial support for European agriculture could be better used. Colitti suggests that it be invested in the expanding markets of industrializing countries.

Whether or not the economic feedback mechanisms predicted by economic theory will work smoothly in practice to pump rich economies from expanding poor economies, it is clear that markets in the newly industrializing and the

---

[5] Colitti suggests that foreign direct investment in industrializing countries is the only way out of the stagnation that many countries are experiencing. Other participants argued that instead of necessarily using technology to expand business and ventures along existing patterns of development, it could be used to clean up the mess that already exists. Colitti's reply is that technological innovation – whatever its use – needs steady improvements which happen much more if an economy is growing.

industrializing countries *are* expanding. There is scope for trading capital stock that is more environmentally benign than that used in the OECD's industrialization. The opposite possibility is that outdated and less efficient capital stock will be transferred. An example was cited of an old UK power station being dismantled and sent to China – but this was claimed to be an exception.

Trade The interaction of environmental standards and trade is recognized as a problematic area, whether the standards are set unilaterally or globally. The case of the US ban on Mexican tuna is now famous: the Mexican fishing nets intended to catch tuna inadvertently trapped dolphins, and the US protection of mammals meant that restrictions were imposed on tuna imports. The fishing process used by the Mexicans was not considered to meet the US's own environmental standards. This highlights the question of whether one country or set of countries should be able to impose its environmental standards on others when countries have different preferences and/or capacities for environmental management. If the production processes that take less care of the environment are cheaper than more 'environmentally friendly' methods, imported goods will challenge the competitiveness of the domestic goods. Do countries' varying preferences and capacities represent further differences between them – like wages, climate and resources – that should be exploited for international free trade using the principles of comparative advantage? Where is the acceptable dividing line between trade protectionism and measures that a country considers to be justified protection of the environment? (Cable)

In addition, application of cost-benefit procedures – which is difficult in any case – becomes increasingly problematic when goods are traded which are part of the cost-benefit valuation process. The cost and/or value that country A puts on an action is open to suspicion on the part of other countries that A is making the valuation with underlying trade protectionist motives. A's export prices could be lowered because it undervalues a domestic resource or product; A could increase the price of imports by claiming country B was undervaluing the environmental externalities of its goods and adding tariffs to B's exports to A.

Protection against protectionism disguised as environmental policies might be necessary: who would define and/or monitor this, and how environmental concerns would be reflected fairly must form central questions in the post-Uruguay Round discussions on trade regimes. International legislation might be the preferred option when dealing with global environmental problems to keep countries from free riding on others' actions.

Technology transfer Transfer and deployment of energy-efficient technology was heralded as potentially bringing great environmental benefits; either through increased energy savings by deployment of efficient technologies or from increased use of renewable resources (an issue focused on in Seiki's paper). Transferring human capital potential through training was also highlighted as being beneficial. There were questions as to:

- Whether technologies should be transferred to less industrialized countries before being extensively tested in richer countries.

- Whether existing trade investment and information structures were sufficient. Seiki's paper suggests that it could be useful to establish a network among supply-side business and/or to set up an intermediary organization providing information on finance and technologies.

  Giving incentives to companies to transfer technology could provide environmental benefits. Seiki questioned whether government subsidies and/or soft loans for transfer of technology could be used to reinforce current patterns of trade.

- Whether transfers would take place more successfully under the auspices of government-led financing – providing low rates of return – or through industrial direct investment. Will industry's 'efficient market solution' indeed be the most efficient and therefore the most attractive option to developing countries, thus negating the need for or take-up of government concessional transfer schemes? Furthermore, it is questionable whether government transfers would be large enough to finance the amount of technology transfer desirable. Industrial direct investment appears likely to outstrip government-supported schemes. Participants seemed in agreement that evidence so far points to this being so, and that industry to industry schemes are often the most effective methods of transferring technologies.

- Whether there is likely to be a significant role for traditional forms of technology transfer. Industrializing countries are generally unwilling, or unable, to pay a premium for clean technologies produced in rich countries. It was argued that poor and industrializing countries will install clean technologies on a larger scale if they are produced in these countries.

The degree of transfer of clean technology will be strongly influenced by whether environmental legislation exists in the industrializing country and the degree to which it is implemented.[6] This will also affect the amount of dirty technologies that get dumped: if clean technologies are readily available there will be fewer incentives to take obsolete technologies that are often more costly to run.

The term technology transfer was used consistently to mean transfer of technology through concessional loans and/or direct investment rather than having any connotation of assistance in the form of grant aid.

International measures to curb environmental degradation Some of the global environmental problems could be solved by an international response. For example, the economics and consequences of assisting the construction of more efficient new energy plants in developing countries versus financing retrofitting technology in industrialized countries could be considered when tackling global environmental problems.

Seiki mentioned more formal innovative approaches, such as emission offsets, as 'joint implementation' or tradeable permits projects, but these were not taken up in discussion.

Energy demand Energy demand and supply in industrializing and industrialized countries differ: often in terms of rapidly expanding energy demand versus sluggish demand and overcapacity. As markets move, developing in some places, contracting in others, energy systems need to be able to respond efficiently. Uptake of technologies depends on the specific energy system in place. Whereas coal use is declining in Europe, for example, it is growing in

---

[6] There will, as always, be 'win–win' situations where clean technology will be the investment of choice.

Asia. The development of clean, efficient coal technology in industrialized countries could greatly affect the amount of such technology installed in industrializing countries. (Baker and Whitehead)

Demand-side management and energy efficiency measures will have different repercussions depending on the state of a country's energy system and the percentage of existing capacity being used. Where new plant would otherwise be needed to meet demand, the savings from not building should be subtracted from any costs of restructuring. In industrialized countries, however, where the system often already has excess capacity, savings in investment are much less immediate.

Obstacles to change

If there are gains to be had, why are they not being taken? Obstacles to change include responding to and working with uncertainties, defensive, negative players, and some groups' perceptions that they have been pushed outside the decisionmaking process.

Uncertainties

Science Scientific issues were debated. One (continental European) participant commented that Anglo-Saxons have an apparent need to continually question scientific findings and at least one of the participants (Anglo-Saxon) doubted the scientific evidence of global warming, its effects and its links to carbon dioxide emissions and climate change.

In the case of global warming, suggesting that decisionmakers wait until the science is proven is not feasible. Decisions are being made now – even if they are to take no action – on the information available now. Complete assurance regarding the effects and links of environmental problems and solutions might not be apparent for many decades. Many would argue that science, in any case, can rarely assure absolute certainty. Waiting for much greater certainty on the science before taking responsive action on environmental problems risks incurring a rise in total costs.

Participants referred to the precautionary principle: its interpretation is not straightforward. In the Framework Convention on Climate Change the precautionary principle was adopted for actions that could 'anticipate, prevent or minimize the causes of climate change'.[7] The Convention states that 'lack of full scientific certainty should not be used as a reason for postponing such measures, taking into account that policies and measures to deal with climate change should be cost-effective so as to ensure global benefits at the lowest possible cost'.

Risk and strategic choice Perceptions and analysis of risk are related to concerns about science and responding in the face of uncertainty. Risk analysis influences how industries choose to diversify their activities; and although energy industries – particularly the oil industry – are thought of as risk takers, they like to calculate the risk to which they might be exposed.

Analysis of risk is related to strategic choice and profitability. Some industrial representatives said they had cut investment in development of clean and renewable technologies because this was not – or not sufficiently – profitable. Other companies have withdrawn from such activities as capital is limited and parts of the world have opened up to them that had previously been closed. Choices had to be made and risks of losses to be weighed against potential profits. Even having considered the risks – which some might think untenable – of drilling in Azerbaijan, some found these preferable to investing in West European and US windpower. Some industrialists argued that by staying with traditional fuels and energy technologies, companies were doing what they did best, thus using funds most wisely. Other participants continued to query the logic and values of the analysis being used, and the choices and actions resulting from such analysis.

Firms working in a competitive market cannot take the long strategic view which is possible for governments. If society wants different decisions to be made by industry, policies and regulations need to be in place to influence industry's decisionmaking, options and actions.

Costs and benefits Although businesses are used to measuring their internal costs, often accurately and consistently, some methods of project analysis still

---

[7] *The United Nations Framework Convention on Climate Change*, Rio de Janeiro, June 1992.

have their problems, for example, not including externalities. (See *Assumptions* below.) Furthermore, measuring benefits is more difficult than measuring costs. What do we measure any savings against? How can we place a monetary value on a clearer river or a healthier woodland? In the cases of energy efficiency and demand-side management, where there are potential benefits, Weinberg makes the important observation that 'electrons can be counted, the lack of electrons cannot' – hence the regulatory complexity of DSM.

Inertia, defensive playing and entrenched positions
Those who see themselves as potential 'losers' may well drag their feet or negotiate negatively in an effort to minimize their losses. Slow action might mean higher costs in the long run. Taking different strategies could better minimize these losses, or in some cases turn them into profits.

Mindsets that are in danger of becoming entrenched should be challenged. There was concern that at international meetings developing countries often behave as if environmental problems represent a means of extracting finance from rich countries; they fail to realize their own vulnerability. Similarly, concern was expressed that rich countries fail to recognize their responsibility for a great part of the environmental problems and their obligation to remedy this. All countries are aware that international markets and activities are changing and that within these changes are great opportunities for developing along sustainable routes.

Insiders and outsiders
Some who fear they could be losers, and who therefore are playing defensively, are already on what could be termed the 'inside'. They are well-organized groups, are within established industrial, lobbying or bureaucratic frameworks, and have access to information, politicians and other policymakers. Some of the 'insiders' are potential 'winners', but others, less established and organized – 'outsiders' – could also benefit from changing energy industry structures. Often these outsiders do not have the resources or the access and entrées to decisionmakers of the more established groups. Their voice does not have the same influence as that of the insiders. Politicization would help, and is helping; for example, through forming bigger coalitions. Increased access to information would also be beneficial.

Some companies feel that they have been pushed outside the process of formulating environmental regulation by environmental groups. Especially in the US, companies tend to believe environmental movement lobbies policy-makers and manipulates the press with arguments for environmental protection that do not weigh up sufficiently the costs and benefits of the regulations demanded.

But there are converse views. Some companies are already working with government and legislative bodies, advising and discussing the impact of pending and potential regulations. Moreover, rather than feeling included, many environmental groups actually feel excluded from 'insider' relationships with legislative bodies. Furthermore, some would say that 'the environment' – species, ecosystems, the planet itself – is the ultimate outsider, which is why environmental problems and/or lobby groups are the only way of 'internalizing' its needs.

General observations

'Sustainable development'

One of the ironies of the workshop is that in an attempt to avoid repeating a process which is taking place ad infinitum, the term 'sustainable development' was not discussed explicitly. With hindsight, an agreed statement of its meaning might have been beneficial as a starting point. On the other hand, if the workshop participants had diverging views on the implications of the term then a rigid definition could have restricted discussion and alienated some participants. Opinions possibly diverged over how far issues of energy planning and usage can be divorced from wider social and political questions of resource allocation and rights of freedom of lifestyle.

Sustainable development appeared to be comprehended as being related to the problem that world resources are limited, and that the present level of consumption of these resources, taking into account projections of worldwide growth of industry, population and per capita consumption, is not sustainable: 'not sustainable' meaning that the world – the global biosphere – will not be able indefinitely to tolerate or support such a way of living. This understanding demands action which will provide a situation that is supportable – that is, sustainable development.

Discussion seemed to stem from an understanding of this unspoken baseline definition of sustainable development, and an acceptance that many were changing their value systems from acceptance of non-sustainable exploitation to concern for sustainable development.

No participants commented that this mainstay of discussion was not explicitly discussed. Indeed, communication between participants was effective and extensive. The fact that these people participated in the workshop was an indication that they had concepts of the subject matter and were convinced of its importance – both currently and in the future.

Synergies

A strong theme to emerge from the papers and workshop discussion is that of synergies, in both impacts and responses. Environmental problems can be local, regional and/or global, but the whole can also be greater than the sum of the elements. Things react together: pollutants (Nilsson) and solutions (Weinberg); the environment and economics.

The problems and solutions can be cross-sectional and cross-border. Where global or regional environmental issues are being taken into account, the inter-dependency of responses crosses boundaries – national and regional. Harmon-ization of Western Europe's policies cannot be divorced from considering those of Eastern Europe and of areas further afield.

In a similar vein, environmental strategies cannot be divorced from other policies. This cross relevance emphasizes the need for environmental consid-erations to be incorporated into all ministerial policies, and especially those of finance ministries.

Responsibility for changing energy use falls on many sectors – not only the energy industries. An example given was of a law in Japan: after road accidents, cars have to be repaired within six weeks or must be taken off the road. Regular maintenance check-ups – or even declaring very inefficient vehicles obsolete – could substantially decrease fuel consumption.

At a scientific level, the issue of interaction refers to the chemistry between pollutants and the exacerbating effect one set of emissions can have on the effect of another; this needs to be more fully researched and reflected in the development of company planning and environmental policy.

And at the micro level energy savings, too, can come in packages, and customers often need someone to integrate the options available for such savings. (See *Industrial structures* above.)

On the other hand, some of the solutions to the problem are diverse, depending on its location, and so will not be appropriate for blanket application: industrialized and industrializing countries have different problems, needs and aspirations, as do rural and urban areas within the countries.

Future change

The workshop reflected broad (though not universal) consensus that change will be necessary; that societies' environmental concerns will not diminish; and that the energy industries will play their part in taking responsibility for any unacceptable pollution they cause. Areas that need attention include various assumptions, tax structures, and development of clear, well thought out legislation.

Assumptions

Various assumptions underlying patterns of growth came under the spotlight as being misconstrued or ill founded. For the energy industries to work successfully within a sphere of sustainable development, it was suggested that the following concepts be reconsidered:

- *Equating development with increased consumption of resources* Increased utility, or benefits for a society, can come from the same amount of resources through changing behaviour or using technology in different ways.

  Examples from the US, in particular from California, were provided of how the idea of increased profits coming from increased use of resources could be, and has been, challenged. (Sioshansi, Weinberg) Profits are coming from 'less' rather than 'more'. Electricity supply companies are changing the nature of their business through a mixture of energy efficiency goals and demand-side management. These changes – for example giving free energy-efficient lightbulbs to customers – can appear radical to industries in other areas (both sectoral and geographical). Often initial changes have been made to meet state regulations on limits to plant expansion (indeed alterations in many regulatory systems made some of these changes viable). These industrial changes have developed, however, into being part of the planning strategy of companies who see their role changing to being suppliers of energy services rather than of energy per se. Pacific Gas and Electric has completely done away with its construction division, yet has increased its profits. (von Weizsäcker)

Related to sources of profit is the concern that clean-up activities – for example production of a cleaner fuel – can bring increased revenue and profits but might in the process of providing a cleaner end-product use more fuel than would otherwise have been the case. Whether such activities are a benefit to the environment depends upon the relative value attached to different environmental goals.

• *Price, cost and revenue* Two points were made about linking these terms. The first was that although the political debate tends to focus on the price of goods, it is important to consider the total bill. Environmental taxes or levies by companies in order to fund energy efficiency programmes might increase prices, but those who reduce their consumption may still pay less. (Sioshansi, Weinberg)

The second point was that prices do not take into account the full costs of producing an energy-product – the classical issue of externalities. Price calculations often do not cover environmental costs, including clean-up and the possibilities of spills and other unintentional, environmentally damaging costs. These should be included in the price, along with the traditional costs of a product's research and production and arguably the costs associated with its replacement or substitution. (Tickell)

Consumers usually think of price rises as bad, but prices are practical indicators for the market. In the case of incorporating externalities, prices would be real indicators of the theoretical acceptance of the concept of sustainable development. If society wants to operate within a framework of sustainable development, then higher resource prices – if they are a reflection of limited resources and/or of environmental externalities – cannot be viewed negatively. Von Weizsäcker argues that without a substantial increase in energy prices, profit incentives will not be sufficient to make what he sees as the necessary major changes to achieve sustainable energy systems. He challenges the view that high prices are damaging, and provides a graph which positively correlates energy prices with economic performance. (Von Weizsäcker, Figure 3.) Seiki calls for China to increase its energy prices, arguing that without increases in price to reflect the value of resources including their environmental impact, there is little incentive for industrializing countries to install energy-efficient technologies and processes.

Faced with a supply of goods produced more cheaply in industrializing countries, companies and governments of industrialized countries hesitate to increase domestic product prices (especially by direct taxes) in a bid to reflect environmental externalities. Industrialized countries fear that their goods will become less competitive and be undercut by imports. International legislation on process standards would theoretically protect against different countries having different standards, but even having agreed regulations and standards does not guarantee their enforcement.

• *'Winners and losers'* The question of 'winners and losers' surfaced repeatedly. 'Losers' can be expected to impede policy developments. 'Winners' might promote formal change through legislation; however, if they are profiting from independent, innovative action, in which they gain from moving first, they might be content not to have legislative requirements create competitors for them.

Since the industrial revolution, some have gained more and some have gained less from economic and industrial activities: 'winners and losers'. This has affected people 'horizontally' – rich and poor – and 'vertically' – intergenerationally. It has affected people directly and indirectly. Is it possible confidently to identify who or what falls into which category?

In the energy industries, defining the winners and losers is not easy. Having considered society's prevailing desire for sustainable development and determined the required action, it was argued that the industries and companies that adapt early will benefit most. Japan is seen as having acted first in the development and exportation of energy-efficient products, and thus already having taken the 'first mover's advantage'. This could be a reason for others to drag their heels. Some industrialists commented that they could act too soon; another participant contested this, saying he thought industry was an innovator and that it was not an issue of being too early; others pointed out that it is usually better to be second or third rather than last.

In some cases being an initiator also opens the doors for others to free ride; but innovation is continual. Patenting laws are designed to protect against free riding. Comments highlighted differing views about the adequacy of such protection, and whether it enhances sustainable development (by encouraging innovation) or restrains it (by impeding the transfer of knowledge and technology).

Unless compensation is paid, the obvious losers from limitations on carbon dioxide in particular will be the owners of fossil fuel resources.[8] Countries which receive large amounts of revenue from fossil sources may lose. The extent to which companies owning polluting resources lose depends on where they put their current investment and whether they diversify.

Beyond this instance, however, discussion did not identify the groups that by definition would lose.[9] Being flexible and responsible is necessary: such players will be the 'winners'. Indeed, the importance of attitudes and mindsets became very apparent in the workshop itself, as people from the same sectors of society considered the same issue, but made comments that were very different in tone and conclusions; perhaps those who lose out will be defined more by mindsets than by their current role or activities.

Although they discussed various response scenarios, the industrial participants appeared inclined to think that there would necessarily be 'winners and losers'. This way of thinking could promote early action, as companies attempt to profit from the changing world rather than be the 'losers'.[10]

---

[8] From a broader perspective than that of company profits and competition, one participant underlined the need to take seriously the potential 'losers' in OPEC, warning of the possible social and political unrest attached to loss of oil revenue if earnings were restricted by a decline in global oil demand.

[9] Von Weizsäcker does identify specific groups which would lose if his suggested 'ecological tax reforms' were undertaken. See *Tax structures* below.

[10] See also *Insiders and outsiders* above.

Tax structures

Von Weizsäcker and Colitti maintain that subsidies on a range of non-sustainable activities persist even in OECD countries and that abolishing them could be beneficial. Von Weizsäcker extends this by arguing that high taxes should be phased in over time on non-sustainable activities: for example, use of non-renewable fuels and certain chemicals, and certain types of land use. He notes that a 5 per cent annual increase in prices for 'problematic factors' produces a doubling of prices in 14 years and an eight-fold increase in 42 years. Setting a long-term plan of tax rises would encourage productivity gains which would greatly lessen the actual impact on high street consumers. Applying such a tax to energy, allowing for productivity gains and a modest share of energy in total costs, von Weizsäcker calculates that consumers would face an overall price increase of 0.08 per cent annually.

Potential 'double dividends' are available from introducing taxes aimed at limiting environmental degradation – taxing 'bads' – while simultaneously reducing other taxes, for example income/labour taxes. A counter argument was that if environmental taxes were successful in abating environmental degradation, in the longer term tax revenue would drop to such a level that it would provide insufficient income for a Treasury. At this point policies would have to be reassessed, shifting taxes elsewhere, or cutting government expenditures. However, this would be decades away from the instigation of such a scheme.

Such taxes would affect countries differently, according to their economic structure. For example, compared with industrializing countries where labour is cheap or more productive, West European countries are inherently less competitive in labour-intensive industries.

Furthermore, European oil industries feel that existing taxes on oil are already exceedingly high, and they are reluctant to accept additional taxes. If the aim is to make customers pay a sum that reflects externalities, then this is already happening. Customers in Japan and Europe are paying several times the production costs of this industry's products. Some of the taxes could be set against environmental costs.

In the limited discussion on revenue from 'environmental taxes' some argued that it could be wise to use the revenue to promote activities that support sustainable development. One suggestion was that industry might have the most relevant knowledge and skills, and thus be better placed than governments to use any additional revenue from increased energy taxes or prices. A further suggestion from von Weizsäcker was that, assuming taxes move from labour to environmental 'bads', the revenue could be split between sellers of energy and those who have to pay for labour goods.

Avoidance of excessive, impractical or inconsistent legislation

Consumers, non-governmental organizations, industry and government have all become active in the field of environmental legislation. All have an interest in the route taken by development, and the legislation that pertains to such development.

The most effective solutions to problems that affect all sections of society have the consensus of the diverse groups in society. Industry and other groups should be consulted about legislation, which needs to be farsighted, transparent and consistent, and legislators should be as aware as possible of the implications and effects the regulations might have. These include escalating costs through underestimating necessary remedial action, the scale of the problem and the legal and administrative drains on the scheme. With clear, well thought out legislation, industry can assess the risks better and plan accordingly: to the point of participants leaving the field if they feel the demands on them are too great. (Mitchell)

Non-industrialized countries have different needs from their industrialized counterparts, not only regarding energy systems and technologies, but also regarding legislation. Poor countries often have a basic problem of not having sufficient qualified people or resources to implement and enforce the initial laws, let alone to respond to unforeseen developments.

Steps forward

The energy industries have always faced risks and challenges. Environmental legislation is providing new ones. Although many view industries as the problem, industries will also have to provide many of the solutions. What is needed is not a technical revolution, but rather a structural change to take advantage of better technologies which are often already in existence, and a new emphasis upon sustainable services. Those who tackle the new challenges with imagination and innovation are most likely to derive success and profits. Those who do not act may turn out to be 'losers': arguably, if the market works efficiently their losses could be seen as 'creative destruction'.

Both market forces and regulation can help movement towards sustainable development. In the US both methods have been used at different times and in different situations, and the experience within the energy industries is that a mixture of methods works best. (Sioshansi) A framework of policy and regulation is needed that uses market forces and business processes and simultaneously moves them onto a new trajectory of continuous improvement. The challenge industries are facing is to play a constructive and vigorous role in shaping the framework.

For energy industries to work in societies where sustainable development is valued, they need to be aware of interactions of many factors and cultures. Many of the topics discussed at the workshop were approached from completely different angles. Some participants saw opportunities and challenges that were difficult but needed to be faced; others seemed to see threats. Stereotyping was seen to be just that: perspectives bore little relation to whether or not a participant was employed in a particular industry. Neither was there a strict correlation with age – although in 20 years' time the decision-makers in industry and government will be those who began their careers with an awareness of many of the environmental issues they would need to address.

The changes that take place may be large. Mentalities that see development as a synonym for expansion need to alter. Profits will be produced from services and servicing, and not from unrestrained extraction, exploitation and consumption of resources.

There are many challenges ahead, which will probably mean much change. The great majority of workshop participants – representing many of those sectors where the change will be instigated – anticipated change and wanted to be prepared for it.

Part 1

The Challenge of Sustainable Development

Introductory Address:
Concepts and Dilemmas of Sustainable Development

Crispin Tickell, Warden, Green College, Oxford, and Convenor of the British Government Panel on Sustainable Development

Economic Stagnation and Sustainable Development

Marcello Colitti, Chairman, EniChem SpA, Milan

Introductory Address:
Concepts and Dilemmas of
Sustainable Development

Crispin Tickell

The industrial revolution, when looked at as a whole, has brought enormous benefits. According to the normal (but highly misleading) definition, economic wealth has risen at an almost incredible rate. Global GNP was around $600 billion in 1900. By 1960 it was $5 trillion and by 1988 $17 trillion. But not only has this growth been highly uneven, it has been achieved at a price. We have created what future generations will surely find a peculiar society, hooked on fossil fuels, and pulled by a consumerist philosophy out of synchrony with the natural world both animate and inanimate. This way of life cannot be sustained, and yet it appears to be inescapable. Let us consider five of its aspects.

First, the growth in human numbers: from less than 10 million 10,000 years ago, to 1 billion in the time of Robert Malthus, to 2 billion in 1930, and to around 5.5 billion today. The human population is now increasing by over 93 million a year.

Second, degradation of land: according to the World Resources Institute as much as 10 per cent of the vegetation bearing surface of the earth is suffering moderate to extreme degradation as a result of our agricultural and industrial activities. Enough good topsoil is lost around the world every year to cover the agricultural regions of France. Sixteen per cent of the landmass of the former Soviet Union was judged an ecological disaster area by the now defunct Soviet Academy of Sciences.

Third, pollution of rivers and seas: urban sewage and industrial effluents are a major threat to rivers and coastal waters. In many areas of the world the quantity has outstripped ability to cope. Pollution of fresh water is all the more serious at a time of rapidly increasing demand. Such demand doubled between 1940 and 1980, and is likely to double again between 1980 and 2000. Even in the remotest seas, such as the high Arctic, significant quantities of such toxic chemicals as polychlorinated biphenyls, DDT, lindane, lead and cadmium have been found in air, seawater, sediments, fish, mammals and seabirds.

Fourth, changes in the chemistry of the atmosphere: acidification, loss of stratospheric ozone, and the increasing concentrations of greenhouse gases. Acidification is local in character, and can be remedied with money and political will. Ozone depletion has global consequences. More important than the danger of increasing human melanomas is the potential effect on other organisms from vegetation and crops to phytoplankton in the ocean. Climate change is the least predictable, and potentially the most serious of all. Global warming enhancing the natural and indispensable greenhouse effect could affect every aspect of human society. The main conclusions of the Intergovernmental Panel on Climate Change (1990, updated in 1992, and to be updated again in 1995) represent a broad scientific consensus. On the assumption that we continue to pump carbon dioxide and other greenhouse gases into the air at current rates, the Panel concluded there could be a rise of global mean temperature of around 1°C by 2025 and around 3°C by the end of the next century. Compare this with a drop of around 5°C during the last glacial episodes. Estimates of sea level rise are less precise but the sea could have risen by around half a metre by the end of the next century. But averages are misleading: effects will be much greater in some areas than others, with major disruptions of weather systems, especially in regions already at climatic risk.

Last, depletion of the diversity of life: over 99 per cent of the species that have ever lived are now extinct. But they did not all die out at once. Humans, as they alter and destroy whole ecosystems, are causing extinctions at up to 1,000 times the normal rate. As E.O. Wilson has written:[1] 'a fifth or more of species of plants and animals could vanish or be doomed to early extinction by the year 2020 unless better efforts are made to save them'. Our understanding of the complexity of ecosystems is very limited and we have little idea which are more important to us than others.

It is too late to prevent all these problems; but over time they can be mitigated and we can adapt ourselves to them. So far we have still to recognize their

---

[1] E.O. Wilson, *The Diversity of Life*, Allen Lane, London, 1993.

gravity, and in particular the nature of the relationship between population, resources and environment. It is hard to convince people that there are limits either to the supply of raw materials or to the sinks for the pollution they cause.

Indeed resources seem to have become more abundant. Between 1948 and 1989 commodity prices fell by almost 45 per cent in relation to manufactured goods, although the environmental costs of industrial agriculture and large scale monocultures have yet to be faced. Greater energy efficiency and conservation still have large potentialities. Industrial countries now use much less energy per unit of GDP than before. Most resources have natural limits even if there is always a possibility of finding substitutes for them. Oil is the best example. Current estimates, which have not changed much in the last few years, suggest that there are around 43 years left at present rate of consumption. But studies look back to times when we numbered only two, three or four billion people; and when the overwhelming mass of humanity was consuming at very modest rates. In a world of 11 or 12 billion people consuming energy at industrial country rates, today's oil reserves would run out in 7 years instead of 43.

Fresh water is another case in point. Many countries are already facing serious problems of supply. The United States has persistent and growing regional problems; but these are as nothing compared to coming shortages in China, still less compared to those in North Africa and the Middle East.

If the crisis in resources is yet to be made manifest, evidence of the limits for sinks for our pollutants is all around us. Bursting landfill sites across the industrial world, transboundary shipment of hazardous wastes, and the increasing prevalence of contamination of the groundwater we depend upon, are all reminders that the capacity of the land to absorb waste products may already be reaching its limits. In the meantime we have little idea of the capacity of rivers and seas to recover from pollution. In Europe the most depressing example is the fate of the Aral Sea. The cumulative cost of cleaning up the legacy of pollution in the United States will be between $400 billion and $1 trillion (depending upon stringency of standards). The cost to the countries of the former Soviet Union and Eastern Europe may be greater still. People have scarcely begun to contemplate the costs of current growth in Southeast Asia. Of course there are enormous possibilities in making better use of land through biotechnology. But this is mostly a feature of life in industrial countries.

The problems caused by depletion of stratospheric ozone are almost universally recognized and are a subject of international agreement. Yet we are still adding to the quantity of ozone destroying substances, and do not know how long it will take to restore natural conditions, nor what the consequences in the meantime will be.

The same is true of our production of greenhouse gases. The carbon cycle is imperfectly understood, and thus the capacity of the earth – whether terrestrial or marine – to absorb the extra carbon we are adding to the atmosphere. The seas absorb a certain proportion, while increased growth of boreal forests has probably been a major sink for most of this century. But felling of large parts of these forests could remove even this temporary safety valve. So far the industrial countries have behaved as if the only issue were the destruction of tropical rain forest. Yet Alaska's official policy is to increase 20-fold the number of trees felled, and Russia's forests are being felled at an increasing rate.

To confront this interlocking and wide-ranging group of issues we need above all to think differently. A good place to start is the 1987 Report of the World Commission on Environment and Development[2]. In this, sustainable development was defined as 'meeting the needs of the present generation without compromising the ability of future generations to meet their needs'. Moving towards intergenerational equity has many implications: among them stabilizing population levels; protecting natural systems; merging environment and economics in decision making at national and international levels; seeing resources as a kind of capital stock; accepting limits to economic growth; and recognizing global interdependence.

Two principles need to be put into practical effect. The first is the precautionary principle or approach, well defined in the Declaration to which all governments committed themselves at the UN Conference on Environment and Development in June 1992: 'where there are threats of serious or irreversible damage, lack of full scientific certainty should not be used as a reason for postponing effective measures to prevent environmental degradation'. The second is the principle that the polluter should pay (adopted by the OECD countries as long ago as 1972).

---

[2] United Nations World Commission on Environment and Development, *Our Common Future*, United Nations, New York, 1987 (also known as 'The Brundtland Commission Report').

It is not easy to give environmental consideration due weight in decision-making. There is a clash of logics. Economic growth, as enunciated by most politicians, is often cited as the only way out of our problems. But growth in this usual sense is most misleading. It takes no account of the impact of growth on the environment nor of the inevitable loss of natural resources. On current reckoning the felling of a forest can be counted as a marketable asset with a plus for the balance of trade. But in development terms it can be a disaster: the conversion of a living asset into a dead desert, and a minus for future generations.

Limits to growth do not imply limits to development. Economic activity has been seen as a circular flow of exchange value on an ever increasing scale. But it also entails the consumption of non-renewable resources and the production of wastes. This is rather as if biology tried to understand animals only in terms of their circulatory systems without recognizing that they had digestive tracts as well.

Thus we should recast our vocabulary and distinguish growth from development: growth refers to quantitative scale, while development is qualitative improvement. We need now to look at the optimum management of our economies within the limits of sources and sinks and on the basis of these principles: the rate of use of renewable resources (soil, water, forests) should not exceed the rate of regeneration; the rate of use of non-renewable resources (fossil fuels, minerals, fossil groundwater) should not exceed the rate at which sustainable alternatives can be developed; the rate of emission of pollutants should not exceed the capacity of the environment to assimilate them.

As has been well said in the paper on Sustainable Development published by the British government on 25 January 1994 'economic development is important to any society, but the benefits of any development must be sufficient to outweigh the costs, including the environmental costs'. Such costs must of course include the costs of not taking action when, as in so many cases, it is necessary to take it.

Costs involve pricing. Prices should always tell the truth. They should reflect three elements: the traditional costs of research, process, production etc.; the costs involved in replacing a resource or substituting for it; and the costs of the associated environmental problems.

Pricing of energy is, for example, bizarre. Current oil prices represent extremely short-term reactions to supply and demand, and ignore almost totally the cost of replacing or substituting for a diminishing resource. The pricing of transport is equally bizarre. It takes almost no account of environmental costs, with all its implications for choice of one means of transport against another, the waste of resources involved in the production process, and the curious devaluation of public transport which is so evident in our societies. No wonder that one of the conclusions of the Panel set up by the US National Academy of Sciences in 1990 and chaired by Senator Evans of the State of Washington, was that:[3] 'On the basis of the principle that the polluter should pay, pricing of energy and use should reflect the full costs of the associated environmental problems'.

Pricing should include the removal of subsidies that encourage unsustainable practices. Such natural resources as water and energy are often underpriced in both rich and poor countries. Water supplied by public agencies for irrigation in parts of the United States is sometimes subsidized by between 85 and 90 per cent. Germany subsidizes the production of coal. Other subsidies are less obvious: for example, price supports for fertilizers, pesticides and seeds, which can in turn distort international trade.

As we move from principles towards actions for sustainable development we should clarify the notion of natural resources as capital stock. Sustainability means living on income rather than capital. Hitherto we have often treated natural capital – the natural world – as if it were income, and have focused on increasing human or manmade capital. This was manageable in the 1950s when the scale of the economy was much less, but that is no longer the case. Until now the UN System of National Accounts has been the standard framework for measuring a country's macroeconomic performance. It is supposed to be integrated, comprehensive and consistent. Yet it is none of these things. It does not treat natural resources like other tangible assets, and it does not include activities that increase or deplete them. The result is to produce a grossly misleading picture of national wealth. Depletion of assets can be presented as income or growth. Thus increasing poverty can be read

---

[3] US Academy of Sciences, *Policy Implications of Greenhouse Warming*, National Academy Press, Washington DC, 1991.

as increasing wealth, and the ability of the country to develop, engage in trade, accept loans, and play a role in the world of economic systems is distorted beyond recognition. Many of the problems which have arisen between industrial and other countries can be traced back to this fundamental intellectual flaw.

The need for new thinking is well recognized. But we need to be clear not just about what is to be done, but about who is to do it. There are three main channels of action: governments, acting together or singly; business; and public participation through education, academia, non-governmental organizations, unions and others.

First, the role of governments. Already they are under a host of obligations. Among others are those arising from the Convention on the Transboundary Movements of Hazardous Waste and their Disposal; the Vienna Convention on Ozone, the Montreal Protocol and subsequent agreements; the Framework Convention on Climate Change, the Biodiversity Convention and Agenda 21 with its requirement to report to the UN Commission for Sustainable Development; the Global Environmental Facility; and the new GATT agreements resulting from the Uruguay Round. Some of these agreements are more useful than others. Many are ambiguous in character.

Most countries have acclaimed the recent GATT agreements. However imperfect, they will serve to increase international trade. But unregulated free trade is no more capable of delivering environmental protection internationally than are unregulated market forces of delivering it nationally. The GATT final act should have written into it the concept of sustainable trade as part of the overall concept of sustainable development, and the GATT Working Party on Trade and the Environment needs new and more balanced terms of reference. Trade as a source of wealth is too important to be left to traders and trade experts.

Governments are likewise under a multiplicity of national obligations. At home they set ground rules through regulation (where level conditions have to be set); market instruments (such as fiscal incentives and disincentives, and tradeable emission permits); promotion of better environmental management; encouragement of new technologies and greater efficiency, particularly in the field of energy; establishment of key environmental indicators; and introduction of environmental accounting in national budgets.

Governments cannot act alone: they need full public participation. Some are already seeking it. The US Administration has established a President's Council on Sustainable Development, which draws together members of the Administration, industrialists and environmentalists to guide government strategy. In Britain three incentives for working together were announced on 25 January 1994: a Government Panel on Sustainable Development (to give advice and monitor environmental policy); a UK Round Table on Sustainable Development (bringing together representatives of the main sectors or groups); and a citizens' environment initiative (applicable at regional and local level). In Britain there is already a system of green Ministers, a network of Ministers, one from each government department responsible for the environmental implications of each department's activities.

The role of business is equally fundamental. All business activity has to take account of environmental considerations. This has been fully recognized by such bodies as the International Chambers of Commerce, the Business Council for Sustainable Development and the World Industry Council on the Environment. Some countries – notably Japan and Germany – are much further ahead than others.

In Britain there is the Advisory Council on Business and the Environment; the Business in the Environment initiative; a new environmental programme for senior executives under the auspices of the Prince of Wales; and initiatives from the Confederation of British Industry and the Trades Union Congress. So far the best response has come from the larger companies.

Put negatively, businesses which do not respond to environmental issues will find it more difficult to sell their products; dispose of waste; keep within the law; obtain insurance; attract finance; and recruit and retain able staff. More positively, good environmental care offers great opportunities, including cost savings (for example industry could reduce its costs by US$1.5 billion if it applied the techniques for effluent and water consumption minimization used successfully in the Aire and Calder Project);[4] an increasing market in environmental goods and services (predicted to rise from US$200 billion today to US$300 billion worldwide by early next century); and competitive advantage from Ecolabels and associated standards.

---

[4] 'HMIP Builds on Yorkshire Waste Reduction Project', *The Ends Report* (Environmental News Data Services), No. 225, October 1993, pp. 5–6.

Public participation is already rising. In Britain there are more members of environmental and conservation groups (some 4.5 million out of a population of nearly 58 million) than there are members of political parties. They amount to a formidable source of persuasion for improvement, but, as in the United States, they tend to push and pull in different directions. There is also a widespread sense of unease, as every public opinion poll shows. People know that something is wrong in the way we run our societies, and in the values we have inherited. But they are uncertain and confused about what to do. Hence the importance, as we attempt to move towards sustainable development, of education in schools and universities: a core curriculum in schools that places sufficient weight on environmental education; the access of non-governmental organisations to policymakers; and the role of citizens as green consumers, householders, volunteers, workers, parents and the rest.

I conclude with three general observations that may be of use as we struggle with the concepts and dilemmas of sustainable development. The first is for scientists. Science is full of surprises. We have to maintain open minds, and adapt as we go along.

The second observation is for governments: they have to work together and establish clear rules for themselves as others. Sometimes the science may be ambiguous. Yet the art of politics is making good judgements on insufficient evidence.

The third is for the rest of us, including business, trying to make a living. We need a better understanding of the environment, and even more to make its care a natural part of our way of life. This means taking a longer perspective, and looking to future as well as present generations. One final encouraging thought. People have become rich making a mess over the last two hundred years. In my judgement they could become even richer still clearing it up over the next two hundred.

Economic Stagnation and Sustainable Development

Marcello Colitti

Abstract

The industrialized countries are undergoing a phase of economic stagnation, mainly due to the market saturation resulting from the decline of consumer demand.

The first priority to be achieved is, therefore, to revive growth and development, by enlarging the market and creating sustainable development in non-industrialized countries, where demand is far from saturation.

The sustainable character of development requires new technologies, in order to be ecologically respectable and to avoid replicating wasteful models.

The technological development work has therefore to be directed to the improvement of the existing processes and to the attainment of a new model, based on three basic objectives: reduction of the use of raw materials and energy per unit of product; reduction of the cost of putting in order the old sites of production; and reduction of the emissions of poisonous gases and effluents.

These fundamental objectives can be better achieved if the industrialized countries move out of the present stagnation and resume reasonable rates of economic growth, though these will inevitably vary from area to area.

I must first of all confess that I have some difficulty in dealing with the subject of the workshop.

How can we talk of sustainable development when there is no development at all? In the industrialized world, global demand has been growing very slowly, if at all, for some time, and the fear is spreading that this may have structural rather than temporary roots. Personally, I believe this fear to be well justified.

The economies of Western Europe, the US and Japan are indeed in a phase of stagnation. We are not in a recession, we are stagnating. The crystal ball gazers talk about an average GDP increase of 1–2 per cent in any of the next five years, and those who suggest a growth of 3 per cent in the US are considered quite optimistic. In a recession, GDP does not increase much even in 'good' years because demand is sluggish or declining outright.

The reasons are not easy to explain, partly because stagnation as an economic phenomenon went off the economists' syllabuses ages ago.

Perhaps the reason is that in the industrialized countries some important factors are putting the brake on demand for consumer goods. The first of them is saturation. A three-car, two-house, five-television-set and countless-gadget family may not particularly want to increase the heap of consumer goods it is supposed to use. Acquiring a new car would actually create the problem of where to put it, and when to use it: the family would end up considering it a nuisance rather than a facility.

The second factor is population. In industrialized countries demand is not increasing strongly, partly because population is growing very little, if at all. This is also a 'saturation' effect. The cost of raising children in these countries is now so high that many people decide to have them later, if at all; this strongly restrains the increase in demand. These two factors may be at the origin of the exhaustion of the long cycle.

Moreover, due to stagnation some categories of workers – especially blue collar workers – may not have the money to increase their expenditure, or even to keep it at the same level as before.

Technological progress has a sinister edge, too. The better the machine tools, the fewer the workers, the lower the total wages paid. The built-in tendency to restrict employment in manufacturing, and therefore demand, has until now been counter-balanced by an increase of employment in services, in which, alas, productivity increases very slowly, if at all. Some people even say that it is downright improper to talk of the productivity of services. Production jobs lost have been offset to some extent by the increase in these white collar service jobs. Industry has become like the American army, with six back-up men for each soldier in the firing line.

However, this cannot go on. Companies which have not adopted the latest labour-saving techniques and equipment have high production costs, making them uncompetitive and preventing them from keeping on the payroll the number of non-productive people they already employ, let alone increasing it. So, the total wages bill may not increase quickly enough.

Moreover, governments are too deep in debt to apply the Keynesian recipe, which may not work even if applied, given the fact that the modern economy is much more rigid than the one Keynes was talking about.

We are experiencing the vicious circle of stagnation: as demand does not grow, we are left with a crisis in the capital goods industry, a very slow increase in productivity and a slowdown of GDP growth. This is the sum total of the factors we have been quoting up to now, the combined effect of the production cycle of capital goods and of the decline of the main items of consumer demand. We could conclude by saying that the great era of the motor car and the sprawling city which supported demand for practically everything is over. The result is that industry, and especially those industries which have been the beneficiaries of such an era, is suffering deeply. It seems that there is no way to solve this problem without tackling the macro issues.

What should we do? Well beyond the scope of any single industrial company, and even of any single industrialized country, at the macro level the only thing to do is to enlarge the market by creating development in non-industrialized countries. This can be done by exporting capital, that is, investing abroad, a measure which has been taken many times in history for exactly the same purpose. Some historians even attributed the survival of capitalism as an economic system to the creation of the great North American market as a result of European, mainly English, investments in the seventeenth and eighteenth centuries.

Just imagine what the OECD countries could do by simply redirecting the $350 billion per year they waste in supporting their own agriculture! Suppose they had invested some of this money in agriculture in Third World countries. How many tonnes of plastics or steel would we now be selling, directly or incorporated in various goods, to, say, Mali, or Vietnam?

This misdirection of investment, I am afraid, is the fundamental problem of the world economy.

We have to resume development. I am well aware that this is a non-economic choice, but a value judgment. A value judgment, let me say, which looks very much like one made by Monsieur de la Palisse. In poor countries, economic development is better than poverty, which means physical pain due to hunger, bad accommodation and sickness, and an early death after a miserable, short life. It is, however, clear to me that development in these countries is not going to accelerate unless there is growth in the developed countries.

Of course, we have just said that the countries that are rich, or over-rich, are stagnating because of market saturation. So we have to create growth of demand somewhere else, supplementing the spontaneous trend of the world economy, which is ensuring that some countries in Asia, which are not poor now, but certainly were poor a few years ago, are experiencing the only real growth left in the world economy.

Capital exports from stagnating countries would almost certainly do the trick, if used to finance investment projects, especially industrial ones, rather than to subsidize consumption. This is a sure way to revive capital goods demand, and consumer demand may very well follow.

Suppose, to make things simpler, we single out an area for investment which does not compete with anything in Europe, but would actually improve the supply opportunities of the old continent: for example, natural gas.

We all know that Europe will need more gas in the next two decades. Its present supply structure looks like a wheel with one spoke missing: the north/south is there, the north-east/south is also there, and so is the south/north – I mean, gas coming to Europe from the North Sea, Russia, and Algeria. What is missing is the south-east/north-west link: gas from the Middle East coming into Europe. And one of the routes could be through North Africa, for example, via a big feeder running along the coast.

In North Africa we have a rather convenient mix of raw materials and investment opportunities, plus the awareness that Europe has to take care of its borders. The southern border spanning the Mediterranean is one of the most sensitive; it has been this throughout history.

North Africa has plenty of gas, and we should take care that it gets transformed into economic development, as was the case, for example, in northern Italy in the 1950s and 1960s.

Gas can be brought to the North African domestic market either as gas or as electricity, as technological changes have made gas-generated electricity cheaper, cleaner and faster to produce than conventional oil- or coal-generated electricity. A gas feeder along the North African coast would integrate local and Middle Eastern gas, and be a big transport as well as a distribution facility.

Natural gas liquids are among the best raw materials for petrochemicals. So we may have the makings of a three-way development programme in North Africa, which would hold out hopes of a fresh bout of economic development there; new gas sources for Europe; and less intractable political problems on our southern border.

I believe that this is a good example of the investment opportunities that can be created in a specific part of the world, which could, at least, increase the demand for capital goods.

I don't mean that investment and electricity in North Africa is going to be enough to steer the world from stagnation to growth. However, the methodology of large projects in areas which need them should be developed further, in order to produce a set of concrete proposals for reaching the goal of international economic growth.

The present is actually a good time for such an enterprise, because stagnation is touching the financial world too, and there is a lot of money floating around that has not been put to good use. We may even find ways and means of financing such plans, partly from private sources, provided the projects are bankable, that is, provided they can be considered eligible for project financing.

What is needed to carry out project financing outside the traditional area in which this technique was born – the development of an oil field – is a set of guarantees from the various partners in the investment. These include a completion guarantee from the technical partners and the chosen engineering company; a raw material delivery agreement, at prices known to and approved by the lending banks; a full lifting agreement from a credible international marketer, covering the full production of the investment; an escrow account through which sales proceeds will flow, and from which the lending banks can extract their dues; a cash deficiency agreement, involving both raw material and finished product prices, in case the market deteriorates markedly; and, finally, a commitment from the host country not to interfere with the work. Clearly, these conditions require the presence of a major company – one that is accepted by the financial institutions, has a consolidated market or can find one, knows how to negotiate with banks and governments, how to put up a feasibility study and, above all, how to distinguish a profitable from a non-profitable project.

We have now to take into account the second item of the equation. We don't simply want development, we want sustainable development.

I take this to mean that development or growth must be ecologically respectable, that is, entail less squandering than in the past of often irreplaceable assets through wasteful production technologies and inept planning.

This is equivalent to saying that we need technologies, not only for growth, but also to make growth more user-friendly. We most emphatically do not want to replicate in other areas an economic development as wasteful as, for example, that of the United States, which took place in a resource-rich country, at a time when 'sustainability' was not properly understood.

I am not talking here of breakthroughs, of new discoveries that would change all our ways of producing and utilizing goods and services. I am discussing something much humbler, the improvement of existing processes and products, the application of new ideas – wherever they come from – to our production system to make it, step by step, look more like the model we all have in mind. In normal times these innovations come in normal times in a steady flow, but this is somewhat limited and thinned out by the stagnation. We need better technologies, and somebody has to provide them.

Where do new technologies and innovations come from? They have basically three origins:

• The research laboratories of large corporations, that is, the commercial corporations working for profit, and non-commercial corporations, like large universities and foundations, which live by spending those same profits.

- The companies already producing goods, which as a matter of course continually aim to improve their products. This applies particularly to the companies producing capital goods, which continuously improve their machines to compete on the market.

- The individual innovator, who may be a former peasant or a humble mechanic, who creates a new product, a new service or a new, resource-saving way to produce existing products or services.

How do we stimulate the improvement of technologies or the appearance of new ones?

The first obvious engine of innovation is economic development itself. A market in full development will produce new protagonists like the above-mentioned humble mechanic; established production companies will be impelled to invest by rising demand. Usually the larger the number of a certain capital good produced, the better its quality. The more a certain process is embodied in industrial plants, the better it becomes, by adding up all the improvements brought in at every edition, as it were.

Development also produces more money for research. It is well known that over time large research establishments tend to increase their expenses and to reduce the rate of applicable discovery per dollar spent, so that the increase in expenditure has to be large to get any results.

The second engine of innovation may be concentration of effort. It may be impossible to direct the animal spirits of the newcomer who succeeds because he is an innovator: he will do whatever he is capable of doing, and trying to direct him would only result in putting a brake on his inventiveness. However, we can direct the efforts of the large corporations, both those involved in production and those in education, which may be persuaded to share in some kind of objective, to further their attainment.

What are the objectives?

We have to redirect the flow of technological innovation towards three basic objectives, all related to the need to reduce the cost of attainment of a better ecological balance between the production of goods and services and its effects on the environment. These objectives are:

- To reduce the use of raw materials and energy per unit of the product or service supplied throughout the economy;

- To reduce the cost of putting in order the old production sites, on which centuries of industry have piled up refuse and poisons, which look extremely costly to clean up; and

- To reduce the emissions of poisonous gases and effluents from vehicles and industrial plant to reduce air pollution in the cities.

The first and the third objectives are closely related; the second is much more important than one might think, especially in developed areas where the industrial sites date back to the previous century.

The objectives I have mentioned are not new, in the sense that some work has already been done, at least on the first and the third, by creating new products, such as the additives for unleaded gasoline, which constitute a very good example of what I meant when I spoke of new technologies. The main additive, methyl-tertiary-butyl-ether (MTBE), has by now a world market of approximately 11 million tonnes per year, which will practically double by the end of the century. When one considers that the first world-scale plant came on stream in 1989, this development is impressive indeed. Interestingly enough, MTBE is produced by world-scale plants which embody what is actually an old technology, dehydrogenation, which had almost gone out of use, but was first revived and then improved, and is now available in two or three completely new versions.

Even more important than the volume of change has been the resistance to change. The US has followed the road of technological improvement of the engines and fuel it utilizes, to the point of indicating in detail which gasoline has to be used in which state. Europe still uses an enormous amount of leaded gasoline, and the European oil companies are desperately opposing the very concept of reformulated gasoline.

We have the technology, the new engines, the new products, which have entered the market with remarkable speed; we need to apply them – as any of us taking a stroll in a European city might be willing to say – but we are set on the slowest possible course for doing so. Instead, we have concentrated our efforts on the limitation of carbon dioxide, and on the greenhouse effect, for which we have no technology, and on which the scientific evidence is in fact beset by doubts and uncertainties. Do I detect some kind of what the Italians call 'una fuga in avanti', that is, a wilful attempt to conceal the resistance to solving one problem by pointing out other, perhaps less solvable, ones?

Let me end by putting forward a paradox: our objectives can be achieved better and faster if we move out of the present stagnation and resume reasonable – if area-differentiated – rates of economic growth.

Part 2

The Context and Constraints

Environmental Challenges to the Energy Industries

Lars J. Nilsson, Research Scientist, Department of Environmental and Energy Systems Studies, Lund University, Sweden, and Visiting Research Fellow at the Center for Energy and Environmental Studies at Princeton University,

and

Thomas B. Johansson, Professor, Department of Environmental and Energy Systems Studies, Lund University, Sweden

Economic Impacts of Environmental Law: the US Experience and its International Relevance

Valerie M. Fogleman, US Attorney, Barlow, Lyde and Gilbert, London

The Impact of Environmental Policy on Trade

Vincent Cable, Head, International Economics Programme, Royal Institute of International Affairs

Environmental Challenges to the Energy Industries

Lars J. Nilsson and Thomas B. Johansson

Abstract

Urban air pollution, acidification and climate change are three major environmental problems resulting from emissions of air pollutants, primarily related to the use of fossil fuels. The effects of these emissions often entail complex, non-linear and poorly understood interactions between different pollutants and between the pollutants and the environment. Here we review these environmental problems with a view to understanding the limits they imply for future energy systems if these are to develop in ways that are compatible with environmental goals: goals such as air quality guidelines, critical load levels and acceptable rates of greenhouse gas emissions. Emissions reductions of the order of 50 to 95 per cent from existing levels for sulphur dioxide (SO_2), nitrogen oxides (NO_x), volatile organic compounds (VOCs) and carbon dioxide (CO_2) are required to meet such goals.

Current efforts and plans to reduce emissions are inadequate to reach present environmental goals. In addition, demands for reducing emissions are likely to become more stringent as more is learned about the impacts on health and the environment, and as it becomes increasingly recognized that further policy decisions must be made in situations where the scientific knowledge is incomplete.

End-of-pipe technical measures, even with full implementation of advanced technologies, will not be sufficient to meet present environmental goals. Strategies that include system level changes are required. Energy efficiency improvements and a shift to renewable sources of fuels and electricity provide technical options for large emissions reductions. The resulting environmental effects of world energy systems could be minor or negligible compared to the effects of continued reliance on conventional sources of energy.

1. Introduction

Many environmental problems result from the extraction, transportation and conversion of fuels, and from the end use of energy. They range in spatial and temporal scales from indoor air quality problems associated with cooking in developing countries, to land degradation from excess biomass harvesting for energy, to the quality of air and water, to changes in the world's climate over decades and centuries. In addition, the environmental problems often entail complex, non-linear and poorly understood interactions between different pollutants and between the pollutants and the environment.

Some environmental and health problems, such as oil-spills and poor indoor air quality, are largely the result of poor practices and/or management and could therefore, technically speaking, be handled relatively easily. Other problems, such as urban air pollution, acid deposition, and climate change, appear to be much more difficult to address. Although we focus on the latter group of problems in this paper, it must be stressed that concerted action is needed also to address the wide range of other problems.

Present knowledge about the sources and effects of air pollutants and about the effects of exposure to combinations of pollutants is incomplete in several areas, and will remain so due to the uncertainties and complexities involved. The problems of acidification, urban air pollution and greenhouse gases have nevertheless been considered sufficiently serious to lead to government action in most countries. The action started earlier this century with efforts to reduce the number of occurrences of smog and local ambient concentration levels of sulphur dioxide (SO_2) in the 1940s and 1950s. Efforts have since expanded to international conventions for emissions reductions.

Growing recognition of the health and ecological impacts of fossil fuel related emissions has led to increasingly stringent demands for change regarding traditional pollutants, such as SO_2, nitrogen oxides (NO_x), carbon monoxide (CO), particles and soot. We now see concerns being raised about, for example, volatile organic compounds (VOCs) and their carcinogenic effects, and the complex role of photochemical oxidants and their precursors in local and global atmospheric chemistry. In the area of climate change, concerns about potentially large, rapid changes in climate, and the risk that greenhouse warming could trigger such changes, have increased as data from Greenland ice cores have been analysed.[1]

---

[1] J.W.C. White, 'Don't touch that dial', *Nature*, Vol. 364, p. 186, 1993.

   These environmental issues have become increasingly more serious as world energy use has increased, in spite of the fact that considerable and sometimes very impressive emission reductions have taken place. Over the last 100 years, world energy use has grown by on average 3 per cent per year to 8,800 million tonnes of oil equivalent (Mtoe) in 1990. Coal, oil and natural gas contributed 26, 32 and 19 per cent respectively in 1990.[2] If energy use continued to grow, for example, to 13,400Mtoe in 2020 as projected in the reference scenario from the World Energy Council, this would considerably aggravate the concerns discussed here.

   Environmental issues are not the only category of concerns related to the energy system.[3] It is also important to consider economic and social development and security aspects in the development of an energy system for the future. Heavy dependence on imported fuels for primary energy supply places a burden on foreign exchange requirements in many countries, as do the capital investment needs of the energy sector. As a result, financial resources are withdrawn from other pressing development needs. World dependence on oil resources concentrated in a small number of locations creates strategic economic interests in oil, leading to political crises, economic vulnerability and military conflicts. Other major security issues are the risks inherent in nuclear power, in particular the long term risks associated with the growing stockpiles of weapons-sensitive reprocessed plutonium and with plutonium in spent fuel.

   A complete review of energy-related environmental and other societal concerns is beyond the scope of this paper. We limit the discussion to some environmental problems caused by emissions of air pollutants, resulting primarily from the use of fossil fuels. Our evaluation suggests that it is in these areas that some of the more stringent demands on the future energy system will arise. The sources and effects of air pollutants covered by international conventions are reviewed in Section 2. These and other emissions result in urban air pollution, acidification and regional air pollution, and climate change. These three major problems, and the prospects for reaching

---

[2] World Energy Council Commission, *Energy for Tomorrow's World*, St Martin's Press, New York, 1993.
[3] J. Goldemberg, T.B. Johansson, A.K.N. Reddy and R.H. Williams, *Energy for a Sustainable World*, Wiley Eastern, New Delhi, 1988.

environmental goals related to them, are reviewed in Sections 3 to 5. The implications for the energy industries, and some important considerations associated with the main alternatives to fossil fuels, are discussed in Sections 6 and 7.

2. Major air pollutants and their effects

Emissions from fossil fuel combustion include a large number of substances and chemical compounds with a range of health and environmental impacts. In the following subsections, the discussion is limited to the compounds covered by the United Nations Economic Commission for Europe (UNECE) Convention on Long-Range Transboundary Air Pollution (LRTAP) which entered into force in 1983 and its protocols, and to carbon dioxide (CO_2), one of the many greenhouse gases covered by the Framework Convention on Climate Change (FCCC) opened for signature at the United Nations Conference on Environment and Development (UNCED) in 1992.

Many other emissions or compounds are subject to national standards, and to national and international guidelines and recommendations. These emissions include various heavy metals (e.g. cadmium and mercury), particles or soot, methane and chlorofluorocarbons (CFCs). Their role in urban and long-range air pollution and climate change is referred to in subsequent Sections where appropriate.

2.1 Sulphur dioxide
Sulphur dioxide is formed in the combustion of sulphurous coals and fuel oils, and in connection with certain industrial processes. More than 90 per cent of the emissions in Europe come from power stations and other combustion of coal and oil.[4] Figure 1 gives emissions by source for former West Germany, the United Kingdom, and Sweden; these three countries have been chosen to illustrate typical trends in emissions, and their sources, in OECD countries.

The sulphur protocol to the LRTAP Convention, agreed at Geneva in 1984 and signed by 20 countries in 1985, commits the signatories to reduce their

---

[4] OECD, *OECD Environmental Data 1993*, Organization for Economic Cooperation and Development, Paris, 1993.

Figure 1 Emissions of SO₂ by source in (a) former West Germany, (b) the United Kingdom and (c) Sweden from 1980 to 1990, '000t/yr

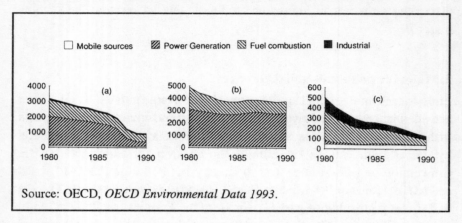

Source: OECD, *OECD Environmental Data 1993.*

SO₂ emissions by 30 per cent from 1980 levels by 1993, a goal which has already been reached and exceeded by several countries. The sulphur protocol is now being renegotiated. The target selected by the Working Group on Strategies of the UNECE LRTAP for the negotiations is a 60 per cent closure of the gap from 1990 emissions to the critical load levels.[5] (See Section 4.)

SO₂ has direct effects both on vegetation and on animals. The effects on vegetation include reduced growth and depend heavily on the simultaneous concentrations of NO_x and ozone. Human uptake is through respiration, causing smarting of the nose and mouth, tears, and exacerbation of asthma, bronchitis and emphysema. The air quality guideline adopted by the World Health Organization (WHO) is that the SO₂ concentration should not exceed 350 micrograms per square metre ($\mu g/m^{3)}$ in any one hour interval and the daily average should not exceed 125μg/m³ on more than 7 days per year.

SO₂ emissions, together with NO_x and ammonia (NH₃) emissions, also lead to acidification of soils and water, and to a range of corrosion damage. The level of soil acidification is determined by the balance between acidifying and

[5] A final draft on the Second Sulphur Protocol was agreed in Geneva in March 1994, to be opened for signature 13-14 June in Oslo.

neutralizing processes in the soil. The ability of the soil to counteract acidification through weathering of minerals – its buffering capacity – depends on the type of soil. In acidification-prone soils, buffering implies that hydrogen ions replace base cations (e.g. Ca^{++}, Mg^{++}, K^+, Na^+).[6] When the base saturation is low, pH falls to a level where aluminium buffering enters, leading to mobilization of aluminium which has toxic effects on organisms.

Soil acidification in combination with high concentrations of SO_2, NO_x and ozone has led to severe forest damage in large parts of Europe.[7] About 20 per cent of Swedish lakes are seriously damaged. Acidification also leads to health risks through increased concentrations of aluminium and copper in drinking water.

2.2 Nitrogen oxides

Nitrogen oxides are formed in combustion processes. Nitrogen monoxide (NO) rapidly oxidizes to nitrogen dioxide (NO_2). Roughly 60 per cent of European NO_x emissions come from mobile sources and the rest are from power generation, other fuel combustion and industrial processes. Figure 2 give emissions by source for former West Germany, the United Kingdom and Sweden.

In the NO_x protocol to the LRTAP convention, adopted in 1988 and now ratified by 20 countries, agreement was reached to freeze NO_x emissions at 1987 levels, or at that of any earlier year, before 1995. In addition, 12 countries undertook to reduce emissions by 30 per cent before 1998. At present, none of the 12 countries appears likely to meet that commitment.

Nitrogen oxides and nitric acid (HNO_3, to which some NO_x is converted) have direct effects on vegetation, especially in combination with other pollutants. Prolonged and repeated human exposure can cause headaches, increase susceptibility to viral infections, irritate the lungs and airways, and increase sensitivity to dust and pollen in asthmatics. The hourly average NO_x concentration should not exceed 190–$320\mu g/m^3$ according to the WHO air quality guideline.

---

[6] A cation is a positively charged ion that is attracted to the negatively charged cathode during electrolysis.

[7] Reports on forest damage are prepared by UNECE, International Cooperative Programme on Assessment and Monitoring of Air Pollution Effects on Forests.

Figure 2 Emissions of NO$_x$ by source in (a) former West Germany, (b) the United Kingdom and (c) Sweden from 1980 to 1990, '000t/yr

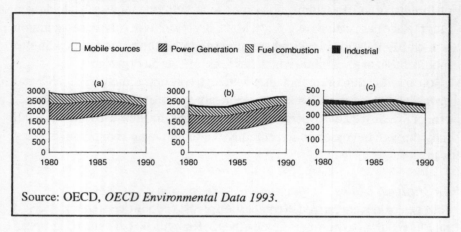

Source: OECD, *OECD Environmental Data 1993*.

NO$_x$ emissions are causing acid deposition and eutrophication[8] of soils and waters. Heavy nitrogen loads also cause leaking of NO$_3$ and the nitrification of groundwater over large areas of northern Europe. The contribution of ammonia (NH$_3$) from agriculture to the total nitrogen load may exceed the contribution from NO$_x$ in these areas. NO$_x$ is also an important precursor to ozone, leading to episodes of ground level ozone formation and global atmospheric change through increasing ozone concentrations in the troposphere and changes in atmospherical oxidation efficiency.[9] Thus, the effects of NO$_x$ emissions and the benefits of emissions reductions should not be considered in isolation.

---

[8] Eutrophication can be defined as the lowering of dissolved oxygen levels in a water system, caused by excessive growth of vegetation, typically algae, due to excessive nutrients, usually nitrates and phosphates, to such a level as to render the system incapable of supporting aerobic forms of life (those living on free oxygen derived from air), including fish.

[9] P. Grennfelt, Ö. Hov and R.G. Derwent, *Second Generation Abatement Strategies for NO$_x$, NH$_3$, SO$_2$ and VOC*, IVL Report B1098, Swedish Environmental Research Institute, Gothenburg, 1993.

Figure 3 Emissions of VOCs by source in (a) former West Germany, (b) the United Kingdom and (c) Sweden from 1980 to 1990, '000t/yr

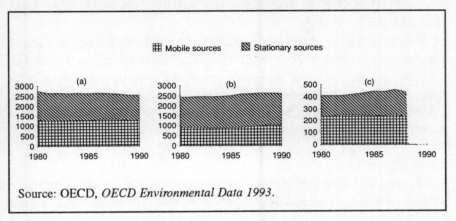

Source: OECD, *OECD Environmental Data 1993.*

2.3 Photochemical oxidants

Ozone and other photochemical oxidants are the result of sunlight-driven reactions with NO_x and VOCs.[10] Mobile sources and solvents appear to account for at least 44 and 37 per cent respectively of anthropogenic VOC emissions in OECD Europe, although there is evidence that the VOC emissions from mobile sources are underestimated.[11] Figure 3 shows VOC emissions by source in former West Germany, the United Kingdom and Sweden.

A protocol on VOCs was adopted in November 1991, the primary aim of which is to reduce the magnitude and number of occurrences of high concentrations of ozone. Most of the signatory countries have committed themselves to reducing VOC emissions from 1988 levels by at least 30 per cent by 1999, although there are some exceptions.

---

[10] Volatile organic compounds include alkanes, alkenes, alcohols, ethers, esters, aldehydes, ketones, aromatics, chlorinated solvents and methane. In the context of urban air quality and vehicle emissions, reference is often made to non-methane organic compounds since methane is almost non-reactive and hence usually does not contribute significantly to ozone formation.
[11] P. Kågeson, *External Costs of Air Pollution: The Case of European Transport*, European Federation for Transport and Environment, Report T&E92/7, Brussels, 1992.

Several VOCs have known toxic or carcinogenic effects, yet effects on health are not addressed in the protocol. The health effects of ozone are similar to those of SO_2 and NO_x. For ozone, air quality guidelines range from 100–200µg/m^3.

Ozone and other oxidants may – due to their oxidizing properties – cause damage to vegetation even at relatively low concentrations, i.e. +50µg/m^3. Crop losses due to ozone have been estimated at 5–10 per cent of total agricultural production in the United States.[12] Elevated ozone concentrations, e.g. +150µg/m^3 during more than 175 hours per year, are thought to be a significant factor in the forest decline observed over large areas of Europe.[13]

2.4 Carbon dioxide

More than half of the anthropogenic contribution to the change in radiative forcing comes from emissions of carbon dioxide. Fossil fuel use in total contributes about half of the change in radiative forcing. Eighty per cent of the CO_2 emissions result from combustion of fossil fuels. The atmospheric concentration of CO_2 has increased exponentially from the preindustrial level of less than 280ppmv to more than 355ppmv today.[14]

Concern about the climate change risks from emissions of CO_2 and other greenhouse gases (GHG, e.g. methane, nitrous oxide (N_2O), CFCs and tropospheric ozone) led to the UN Framework Convention on Climate Change which was signed at the UNCED meeting in Rio de Janeiro in 1992. The aim of the agreement is to stabilize atmospheric concentrations of greenhouse gases at levels that will prevent human activities from interfering dangerously with the global climate system. Some countries have also adopted targets to freeze or reduce emissions by 10–20 per cent from some earlier year over a 10–20 year period as a first step.

---

[12] Swedish Environmental Protection Agency, *Ground Level Ozone and Other Oxidants in the Environment*, Swedish Environmental Protection Agency, Report 4133, Stockholm, 1993 (in Swedish).
[13] R.G. Derwent, G. Grennfelt and Ö. Hov, *Photochemical Oxidants in the Atmosphere*, Nordic Council of Ministers, Report Nord 1991:7, Copenhagen, 1991.
[14] J.T. Houghton, G.J. Jenkins and J.J. Ephraums, *Climate Change: The IPCC Scientific Assessment*, Cambridge University Press, Cambridge, 1990. J.T. Houghton, B.A. Callander and S.K. Varney, *Climate Change 1992: The Supplementary Report to the IPCC Scientific Assessment*, Cambridge University Press, Cambridge, 1992.

Some effects of greenhouse gas emissions have been summarized by the Intergovernmental Panel on Climate Change (IPCC). The change in global mean temperature under scenario IS92a is predicted to be between 0.2°C and 0.5°C per decade.[15]

It is expected that temperatures will increase more over land than over water, and more at higher latitudes than near the Equator. Precipitation is expected to increase on average at high latitudes. The projected rate of change of sea level *due to oceanic thermal expansion* ranges from 2–4cm per decade.[16] Given our incomplete knowledge of climate, we cannot rule out the possibility of surprises.

The effects of climate changes on biota are difficult to predict. For example, changes in CO_2 levels and temperature will affect vegetation directly but will have an even greater indirect effect through, for example, changes in the nutrient balance and moisture levels in soils. It also seems that warming will result in liberation of additional carbon from biota and soils.[17] The effects will certainly be serious for forests that are already under attack from air pollution and acidification.

2.5 Combined effects, complexity and uncertainty

Cause–effect relationships are often difficult to establish both in the case of health risks and of environmental damage. The combined effect of pollutants can be antagonistic, additive, or synergistic. Ozone, NO_x, SO_2 and particulates are thought to have synergistic effects and are known to cause or exacerbate allergies, asthma, and other respiratory conditions.[18]

Direct effects on vegetation from exposure to several air pollutants in controlled experiments are usually additive or synergistic.[19] The combined effects of acidification and exposure to air pollutants on forest damage in

---

[15] The IS92a scenario in the 1992 IPCC supplement is essentially the same as the business-as-usual scenario in the 1990 report.

[16] The emphasis has been added because the estimate does not include potential rapid sea-level rise from large ice-sheets that could be released from polar regions and then melt.

[17] G. Woodwell and F. McKenzie, *Biotic Feedbacks in the Global Climatic System*, Oxford University Press, forthcoming.

[18] Swedish Environmental Protection Agency, op. cit.

[19] L. Skarby, *Air Pollution and Damage on Vegetation*, Swedish Environmental Protection Agency, Report 3049, Stockholm, 1985 (in Swedish).

central Europe are probably synergistic. In addition, trees that are under environmental stress are more susceptible to fire, pests and fungi attacks. In parts of Sweden, there is evidence that forest productivity in terms of stem growth as a result of acidification has *not yet* decreased, due to the simultaneous eutrophication effect from nitrogen deposition. The damage is manifested as needle loss and in a few isolated cases as dying or dead trees. A contributing factor to forest damage in Europe may also be the planting of foreign provenances and species.[20]

Emissions of greenhouse gases (CO_2, CFCs, N_2O, CH_4), acid deposition precursors (SO_2, NO_x, NH_3) and photo-oxidant precursors (non-methane hydrocarbons, NO_x, CO) interact in their influence on stratospheric ozone, warming–cooling of the troposphere and the stratosphere, tropospheric ozone, the tropospheric oxidation efficiency, acid rain and photochemical oxidant formation.[21] The effects of the emissions or changes in emissions are not well or completely understood. For example, current atmospheric chemistry models exhibit substantial differences in their predictions of changes in ozone, in hydroxyl (OH) and in other chemically active gases due to emissions of NO_x, VOCs, CH_4 and CO.[22]

3. Urban air pollution

Emissions in many urban areas, primarily from vehicles, but also from other fuel combustion and other emissions, result in poor air quality. The inefficient domestic burning of coal, charcoal and biomass in developing countries also contributes to poor indoor air quality. Figure 4 indicates the air pollution situation in 20 megacities of the world.[23] Suspended particulate matter (SPM) is a serious problem especially in developing countries. Ozone and its precursors appear to be the most important problem in industrialized countries, while data for these compounds are scarce for developing countries. Only in

---

[20] A provenance is a variation of a species which is adapted to local climate conditions. For example, you can have the same species of spruce growing in northern Sweden and central Europe, but if a specific tree is moved it may not grow well in its new location.

[21] Grennfelt et al, op. cit.

[22] Houghton et al., op. cit., 1992.

[23] United Nations Environment Programme and World Health Organization, *Urban Air Pollution in Megacities of the World*, Blackwell Publishers, Oxford, 1992.

Figure 4 Overview of air quality in 20 megacities based on a subjective assessment of monitoring data and emissions inventories

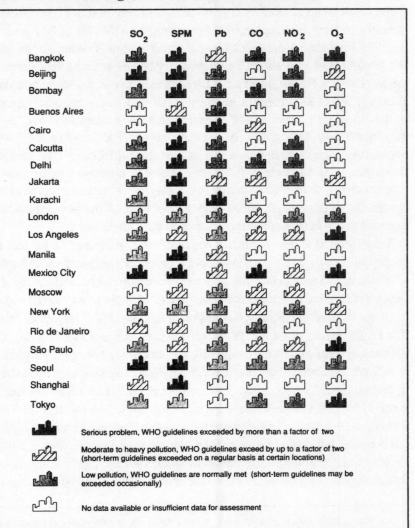

Source: UNEP/WHO, *Urban Air Pollution in Megacities of the World*, Blackwell Publishers, Oxford, 1992.

46 of the 120 matrix elements in Figure 4 are WHO guidelines normally met, although short-term guidelines may still be exceeded.

The result of an air quality classification as presented in Figure 4 depends strongly on the guidelines applied. For example, in the case of NO_x and ozone the WHO guidelines are less stringent than those of Sweden, Japan and the EC. (See Table 1.) For the EC, 135µg/m^3 for NO_x is given as the 98 percentile guide value for 1-hour intervals.[24] Sweden has very stringent standards with 110µg/m^3 as the 98 percentile guide value (though, this is frequently exceeded in locations with heavy traffic where average concentrations over an hour may reach 240–850µg/m^3). At present, about 3 per cent of the Swedish population is exposed to levels that exceed the guidelines.[25] The Institute of Environmental Medicine (IEM) has recently proposed to the Swedish Environmental Protection Agency that 100µg/m^3 should be adopted as the 99 percentile guide value.[26] More than 10 per cent of the population is now exposed to levels that exceed the proposed guideline.

Many hazardous pollutants are not regulated, measured or subject to air quality guidelines. Little is known about sources and concentrations of different VOCs in urban areas since measurements are both technically difficult and expensive. Ozone concentrations are generally taken as an overall indicator of air quality and the increased attention given to VOCs is largely due to their role as precursors of ozone. However, several VOCs also affect the central nervous system or are carcinogenic, for example, benzene, ethylene and butadiene. In general, there are no air quality guidelines for such hydrocarbons in ambient air. The so-called 'low risk level' (i.e. less than one cancer in 100,000 for continuous lifetime exposure) proposed by the IEM in Sweden for benzene is 1.3µg/m^3, for ethylene 0.6µg/m^3 and for butadiene 0.05–0.2µg/m^3. For benzene, which was measured in 17 locations in Sweden during 1992–93, the six-month average concentration in Stockholm and Gothenburg was more than 6.5µg/m^3, or 5 times the 'low risk level'; this was also the case in small

---

[24] The 98 percentile means that the guide value should not be exceeded during more than 2 per cent of the time, i.e. 175 hours per year.

[25] Swedish Environmental Protection Agency, *Pollution and Health*, Swedish Environmental Protection Agency, Report 4139, Stockholm, 1993 (in Swedish).

[26] Personal communication from K. Viktorin, Swedish Institute for Environmental Medicine, Stockholm, October 1993.

Table 1 Air quality guidelines and standards

Guideline/ standard	Time weighted average, $\mu g/m^3$				
	SO_2/1h	CO/8h	NO_x/1h	Ozone/1h	SPM/24h
WHO	350	10,000	190–320	100–200	70–230
Thailand	–	20,000	320	200	100
Philippines	850	10,000	190	120	180
California	715	10,000	470	180	30
US Federal	–	10,000	–	235	50
EC	–	–	135	–	250
Japan	286	–	–	120	100
Sweden	200	6e3	110	120	90

Notes: SO_2: The SO_2 value for Sweden is a 98 percentile value. Percentiles (see footnote 23) are not specified for the other categories. UNEP/WHO does not report hourly values for Thailand, US Federal and the EC. The 24-hour averages reported for these categories are 300, 250–350 and 365 $\mu g/m^3$ respectively.

CO: UNEP/WHO does not list a CO value for the EC. The CO value for Japan is 23,000 for 1 hour.

NO_x: The UNEP/WHO NO_x value should not be exceeded more than once per month. For Sweden and the EC NO_x values are 98 percentile guide values. The value reported by UNEP/WHO for Japan is 75–113$\mu g/m^3$ for 24 hours and for US Federal it is 100$\mu g/m^3$ for 1 year.

Ozone: The US Federal and California values for ozone should not be exceeded more than once per year, and the Swedish value not more than once per month. UNEP/WHO reports no ozone value for the EC.

SPM: The SPM values are difficult to compare due to various definitions used. The low WHO value is for thoracic particles (i.e. particles less than 10 μm in diameter, or PM10) and the high value for total suspended particulate matter (TSP). Values for Thailand and the Philippines refer to TSP. Japan, US Federal and California values are for PM10. Those for the EC and Sweden are 98 percentile values for soot/black smoke.

Source: Data from UNEP/WHO except for Sweden where data are from personal communication with K. Viktorin, Institute for Environmental Medicine, Stockholm.

towns such as Luleå and Boden in the north.[27] For comparison, annual average benzene concentrations at different locations in California vary between 4.1 and 14 µg/m³.[28]

Perhaps the most stringent abatement measures so far have been taken in the state of California. Air pollution levels in Southern California violate federal standards on half the days in each year in spite of Californians already having been relatively successful in reducing emissions, for example by mandating catalytic converters in new cars for more than 15 years. Computer projections indicate that to reduce pollution in Los Angeles in order to meet health standards requires emissions of hydrocarbons, NO_x and SO_2 to be reduced by 80, 70 and 60 per cent, respectively.[29] Even further emissions reductions would be required if California were to meet Swedish air quality guidelines. Roughly half of the hydrocarbon and NO_x emissions in California come from cars. According to current plans in the state, (see Figure 5), the fleet average for cars that are sold in 2001 (i.e. only low, ultra-low and zero emission vehicles) would emit 82 per cent less hydrocarbons and 53 per cent less NO_x than was average for cars sold in 1991.[30]

It is conceivable that the low emission vehicle (LEV) standard can be met with high fuel efficiency internal combustion engine vehicles equipped with catalytic converters (which in turn emit N_2O, a greenhouse gas). However, even a complete shift to LEVs may not be enough to bring about sufficient emissions reductions given the continuous increase in driving. Increased ride sharing, shifts to other transport modes, and the introduction of engine/fuel systems changes, such as electric or fuel-cell vehicles are probably necessary to meet present air quality ambitions in Los Angeles and in many other urban areas. Electric and fuel cell vehicles also have the potential to reduce overall greenhouse gas emissions: see Figure 6.

---

[27] Swedish Environmental Research Institute, *Concentrations of Sulphur Dioxide, Soot and Nitrogen Dioxide in Swedish Urban Areas, Winter 1992–3*, Swedish Environmental Research Institute, Gothenburg, 31 August 1993 (in Swedish).

[28] Personal communication with L. Dolislager, California Air-Resources Board, Sacramento, CA, October 1993.

[29] J.M. Lents and W.J. Kelly, 'Clearing the Air in Los Angeles', *Scientific American*, Vol. 269, No. 4, October 1993, pp. 32–39.

[30] Personal communication from D. Gordon, Union of Concerned Scientists, CA LEV Programme, Berkeley, CA, October 1993.

Figure 5 New car emissions standards in California

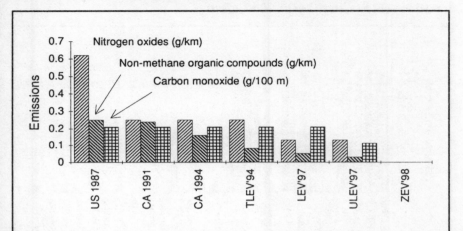

Notes: TLEV = Transitional low emission vehicle; LEV = Low emission vehicle; ULEV = Ultra low emission vehicle; ZEV = Zero emission vehicle. Numbers indicate year of introduction for each vehicle type. For example, in the year 2001, LEV, ULEV and ZEV are projected to account for 90, 5 and 5 per cent, respectively, of new car sales.
Source: Personal communication from D. Gordon, 1993. (See footnote 30)

4. Acidification and air pollution

Southern Scotland was one of the first areas in the world where long range transportation of pollutants was shown to result in large scale acidification. The acidification in combination with air pollution has caused forest damage in parts of Europe and some forest dieback, notably in Poland and the Czech Republic. Southern Sweden, Norway, Scotland, and large parts of Germany and the Netherlands are hard hit by acidification of surface waters, often with high concentrations of aluminium. More than half of the lakes and streams in these areas are seriously damaged with resulting losses in biodiversity and productivity.

Within Europe there are regional differences in the environmental impact of sulphur and nitrogen deposition and ozone levels. In northern Europe a reduction in sulphur deposition loads is particularly necessary. In central

Figure 6 Emissions from alternative-fuel light-duty vehicles relative to gasoline vehicles, in the year 2000, g/km

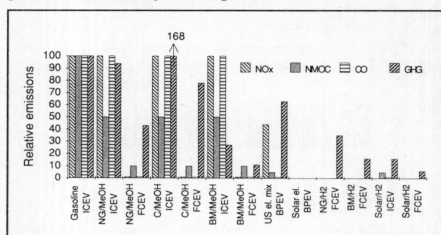

Notes: Baseline emissions in g/km are: NO_x=0.28, NMOC=0.48, CO=3.81 and GHG=282. Emissions are from the entire fuel production and use cycle. NMOC=non-methane organic compounds (virtually the same as VOC depending on definition used); GHG=greenhouse gases; ICEV=internal combustion engine vehicles; FCEV=fuel cell engine vehicles; BPEV=battery powered electric vehicle; NG/MeOH=methanol from natural gas; C/MeOH=methanol from coal; BM/MeOH=methanol from biomass; NG/H2=hydrogen from natural gas; BM/H2=hydrogen from biomass; Solar/H2=electrolytic hydrogen from solar cells.

Source: M. DeLuchi, 'Hydrogen Fuel Cell Vehicles', Institute of Transportation Studies, University of California at Davis, forthcoming 1994.

Europe reductions in sulphur and oxidized and reduced nitrogen are required to reduce acidification and eutrophication. In central and southern Europe reductions in NO_x and VOC emissions are required to reduce ozone exposure levels.[31]

Forest damage is also found in some areas in the eastern part of Canada and the United States, and locally in China, Brazil and Venezuela.[32] In Asia and

---

[31] P. Grennfelt et al., op. cit.

[32] H. Rodhe and R. Herrera (eds.), *Acidification in Tropical Countries*, J.Wiley & Sons, Chichester, UK, 1988.

South America the acidification situation is expected to worsen considerably with current and projected rates of increase in primary energy use. For example, Southeast Asia is projected to experience the same development of sulphur emissions and large scale acidification as Europe, even if emission controls are used.[33] Emissions of NO_x are likely to aggravate the problem.

4.1 Critical load

How much pollution can the environment take? Various attempts to answer this question have led to the formulation of the critical load concept. There are several definitions of critical load in use. One commonly used within the LRTAP Convention has been: 'a quantitative estimate of an exposure to one or more pollutants below which significant harmful effects on specified sensitive elements of the environment do not occur according to present knowledge.'[34]

However, the critical load concept does not accommodate surprises or erratic behaviour in natural systems. An expanded approach is that policy should be guided by the precautionary principle so that lack of full scientific certainty is not used as a reason for postponing measures to prevent or minimize the causes of damage. The precautionary approach was agreed on at UNCED.[35]

Figure 7 shows a five percentile critical load map for sulphur deposition in Europe, based on the Regional Acidification Information and Simulation (RAINS) Model.[36] In a five percentile map, the critical load level is based on what the least sensitive 95 per cent of the ecosystems in each cell can tolerate. Figure 8 shows by how much the critical load level for sulphur deposition is projected to be exceeded in the year 2000 with current plans for emission reductions.

---

[33] H. Rodhe, J. Galloway and D. Zhao, 'Acidification in Southeast Asia – Prospects for the Coming Decades', *Ambio*, Vol. 21, No. 2, pp. 148–150, 1992.
[34] UNECE, *Conclusions and Recommendations of the Workshops on Critical Levels for Forests, Crops and Materials and on Critical Loads for Sulphur and Nitrogen*, UNECE, Document No. EB.AIR/R.30, Geneva, 1988.
[35] For example, see Principle 15 in the Rio Declaration on Environment and Development, United Nations Conference on Environment and Development, Rio de Janeiro, Brazil, 3–14 June 1992.
[36] G. Klaassen, M. Amann and W. Schöpp, *Strategies for Reducing Sulphur Dioxide Emissions in Europe Based on Critical Sulphur Deposition Values*, International Institute of Applied Systems Analysis, Laxenburg, Austria, 1992.

Figure 7 The 5 percentile critical load values for sulphur deposition

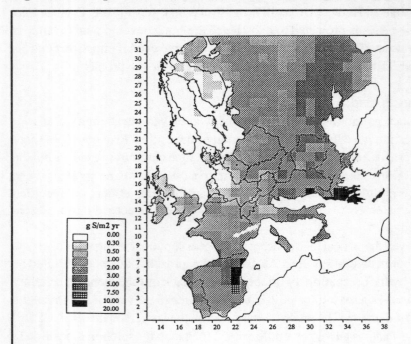

Source: Klaassen et al., *Strategies for Reducing Sulphur Dioxide Emissions in Europe Based on Critical Sulphur Deposition Values*, 1992, (see footnote 36).

Nitrogen causes both eutrophication and acidification. This dual effect make critical loads more difficult to define than, for example, in the case of sulphur alone. Based on a mass balance approach, accounting for nitrogen uptake by vegetation, the critical load for nitrogen on forest soils has been calculated to be between 7 and 20 kilograms of nitrogen per hectare per year (kgN/ha/yr). For southern Sweden the critical load is 6–10kgN/ha/yr. In unmanaged forests with no harvest of biomass, the critical load is 2–5kgN/ha/yr. Current deposition rates over central Europe and southern Sweden are 30–40kgN/ha/yr and 20–30kgN/ha/yr, respectively.[37]

[37] P. Grennfelt and E. Thörnelöf (eds.), *Critical Loads for Nitrogen*, UNECE Workshop at Lökeberg, Sweden, The Nordic Council of Ministers, Report No. 1992:41, Copenhagen, 1992.

Figure 8 Percentage of ecosystems where the critical load for sulphur will be exceeded in 2000, with current plans for emission reductions

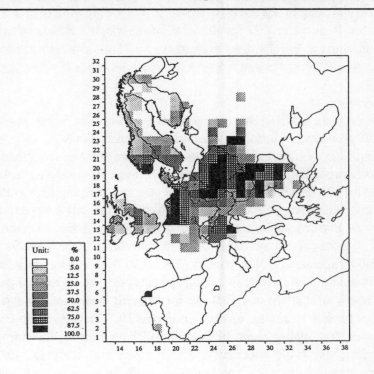

Notes: Current reduction plans for the year 2000 relative to 1980 envisage a reduction exceeding 80 per cent in Germany, Sweden, Finland and Austria; 60–80 per cent in the Netherlands, France and Denmark; 40–60 per cent in Luxembourg, Ukraine, Switzerland, Norway, Bulgaria, Belgium, Italy and the United Kingdom; 30–40 per cent in Belarus, Russia, Poland, Spain, Hungary, Moldova, the Baltic States and Slovakia and the Czech Republic. No reductions or increased emissions are projected for Ireland, Portugal, the former Yugoslavia, Albania, Romania, Greece and Turkey.

Source: Klaassen et al., *Strategies for Reducing Sulphur Dioxide Emissions in Europe Based on Critical Sulphur Deposition Values*, 1992, (see footnote 36).

Critical levels for the protection of sensitive plants and ecosystems against ozone as a single pollutant have been estimated at $150\mu g/m^3$ for a one-hour exposure and $50\mu g/m^3$ for average exposure over the growing period. These critical levels are exceeded over almost all of Europe. It is estimated that the NO_x emissions in Europe need to decrease by 75–80 per cent in order to avoid crop and forest damage in Sweden due to high ozone concentrations.[38]

4.2 Abatement

The prospects for reducing NO_x emissions from mobile sources and experience with this problem in California were briefly discussed in Section 3. While experience shows that it is relatively easier to reduce SO_2 emissions than NO_x emissions through technical measures, e.g. shifting to low-sulphur fuels and adding flue gas desulphurization equipment, even applying best available technology does not appear to be sufficient to reach currently adopted critical load levels. Figure 9 shows a scenario of what could be achieved in the year 2000 if all technical emission control options considered in the RAINS model were fully implemented. The scenario assumes official energy projections and does not account for fuel switching and energy conservation. Modelling results also show that the marginal cost for reducing emissions through end-of-pipe measures increases rapidly: see Figure 10.[39] Thus, to reach critical load levels for sulphur alone in more than 95 per cent of the European ecosystems, one would either have to accept very high marginal costs for reducing emissions from coal and oil fired combustion, or move to system level changes where natural gas, renewable fuels and energy efficiency played a much larger role than they do today.

For power generation in Germany, the present emission standard for a new coal-fired power plant greater than 300 megawatts (MW) is $400mgSO_2/m^3$ (milligrams per cubic metre of sulphur dioxide), 85 per cent desulphurization, and $200mgNO_x/m^3$.[40] This is one of the strictest standards in the world at

---

[38] Swedish Environmental Protection Agency, *Ground Level Ozone*, op. cit.

[39] UNECE, *Integrated Assessment Modelling*, UNECE, Executive Body for the Convention on Long-Range Transboundary Air Pollution, Document No. EB.AIR/WG.5/R.40, Geneva, 1993.

[40] H.N. Soud, *Emission Standards Handbook: Air Pollution Standards for Coal-Fired Power Plants*, IEA Coal Research, Report No. IEACR/43, London, 1991.

Figure 9 Percentage of ecosystems above the 5 percentile critical load for sulphur in 2000, with maximum technically feasible reductions

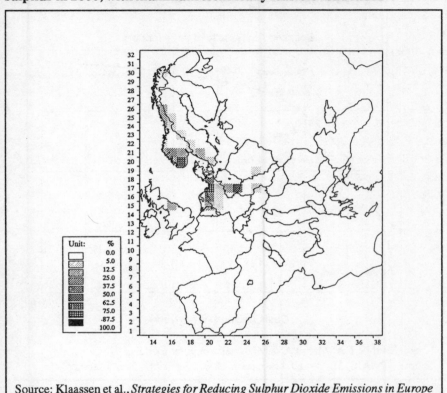

Source: Klaassen et al., *Strategies for Reducing Sulphur Dioxide Emissions in Europe Based on Critical Sulphur Deposition Values*, 1992, (see footnote 36).

present. In an advanced integrated gasification combined-cycle coal-fired power plant, emissions can be reduced to $60mgSO_2/m^3$ if low sulphur coal (approximately 0.4 per cent sulphur) is used and more than 90 per cent of the sulphur is removed. NO_x emissions would be about $300mg/m^3$ and could be reduced to $75mgNO_x/m^3$ if advanced gas turbine technology were used,[41] and even further with selective catalytic reduction. These data indicate that

---

[41] Electric Power Research Institute, *Technical Assessment Guide*, EPRI, Palo Alto, CA, 1989.

Figure 10 Percentage of ecosystems protected as a function of annual expenditure on sulphur emissions reductions

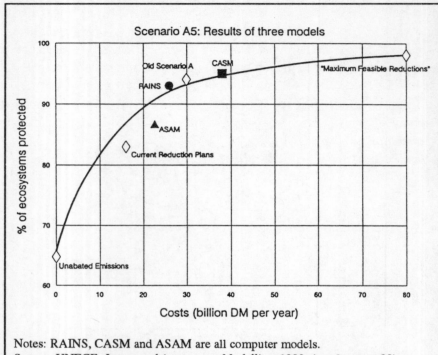

Notes: RAINS, CASM and ASAM are all computer models.
Source: UNECE, *Integrated Assessment Modelling*, 1993, (see footnote 39).

advanced coal combustion technology could possibly produce reductions in the share of SO_2 and NO_x emissions that come from stationary sources to a level which is acceptable by some environmental criteria.

However, advanced technologies for natural gas are much more promising. For example, with natural gas-fired fuel cells the total emissions of SO_2, NO_x, and hydrocarbons would be below $10mg/m^3$.[42] In addition, natural gas when compared with coal generates a smaller quantity of ashes and other wastes, and less CO_2 per unit of energy.

---

[42] Ibid.

5. Climate change

Atmospheric concentrations of greenhouse gases are increasing, and this in turn leads to enhanced radiative forcing. There is scientific consensus that the global mean surface air temperature has increased by 0.3–0.6°C over the last century.[43] There is virtually unanimous agreement that increased concentrations of greenhouse gases in the atmosphere will result in global warming. The point of debate is what the rate of warming will be, how important different climate feedbacks from clouds, oceans, and biota are, and how warming will lead to changes in other climate parameters, such as precipitation and ocean circulation.

As noted in Section 2.5, the gases and aerosols (airborne particles) that directly or indirectly affect the climate include many compounds that are also important urban and regional air pollutants. For example, increased concentrations in tropospheric ozone contribute to the increase in radiative forcing. The cooling effect of aerosols resulting from SO_2 emissions may have offset more than one-third of the heating due to anthropogenic greenhouse gas emissions in the Northern Hemisphere.[44]

The IPCC has predicted rapid increases in global mean temperatures if present trends in greenhouse gas emissions persist. Figure 11 shows the predicted rate of increase in the IPCC scenario IS92a in comparison with variations in air temperature in China since 1380. The burning of fossil fuels now contributes about 80 per cent of the observed increases in atmospheric CO_2, the greenhouse gas which accounts for more than half of the increase in radiative forcing. IPCC estimated in their 1990 report that to stabilize atmospheric concentrations of CO_2 would require emissions to be reduced by at least 60 per cent.

The best estimate for global fossil fuel CO_2 emissions in 1990 is 6.0±0.5 gigatonnes of carbon (GtC).[45] On a global basis this corresponds to roughly 1 tonne of carbon per capita. Stabilizing atmospheric CO_2 concentrations implies allowable per capita emissions of about 500kg carbon per capita per year, which is higher than present per capita emissions in many

---

[43] Houghton et al., *Climate Change*, op. cit.
[44] Houghton et al., *Climate Change*, op. cit.
[45] Ibid.

Figure 11 Variations in air temperature in (a) east China and (b) north China based on documentary evidence

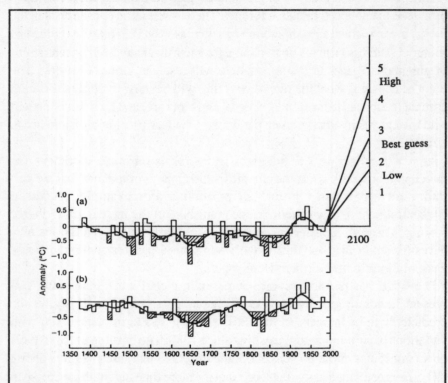

Notes: The smoothed curve is a 50-year running average. The increase in global mean temperature under IPCC scenario IS92a with high, best guess and low climate sensitivity has been added to illustrate the rate of the predicted change.

Source: Adapted from J.T. Houghton et al., *Climate Change 1992*, CUP, Cambridge, 1992.

developing countries. For comparison, the 1990 per capita emissions in former West Germany, the United Kingdom and Sweden were 3.1, 2.9 and 1.8 tonnes per capita, respectively. Strategies to reduce the rate of increase in atmospheric CO_2 concentrations include sequestering, recovery and removal, a shift to low-carbon fossil fuels, a shift to non-fossil fuels, and increasing energy efficiency.

Sequestering carbon through afforestation or reforestation requires large land areas. In addition, reforestation does not provide an infinite sink for carbon. The sink lasts only as long as the trees stand, or are preserved (e.g. pickled) in a way that prevents the carbon being released again. To sequester the CO_2 emissions from a 1,000MW coal-fired power plant requires more than 2,500km^2 of planted forest.[46]

To sequester 3Gt of carbon per year would require an area half the size of Brazil or half of continental Europe to be forested. The area requirement obviously limits potential for sequestering. Using biomass for energy in a sustainable manner would be more effective to prevent CO_2 build-up than sequestering carbon by growing more trees.[47]

For CO_2 recovery and removal, the storage options that have been suggested include piping the gas deep into the ocean or into depleted natural gas reservoirs.[48] Cost estimates are highly uncertain and the technology can only be applied to large point sources which in addition must be located close to deep oceans or reservoirs. An analysis made by the Electric Power Research Institute, in California shows that recovering 90 per cent of the CO_2 in a coal-fired power plant and transporting it 500km by pipeline for ocean disposal would increase the cost of power by a factor of 2 to 2.6.[49] Furthermore, no one knows the effects of ocean storage on the ecology of the sea or how long the gas would remain there or in gas reservoirs.

CO_2 emissions can be reduced from about 0.25kgC/kWh in new coal-fired power plants with conventional steam-turbine technology to 0.2kgC/kWh in power plants using integrated coal-gasification technology.[50] A gradual shift to natural gas, which has a lower carbon content per unit of energy than coal

---

[46] This is assuming that 300 grammes of carbon are released per kilowatt-hour produced, a 60 per cent load factor, and an average assimilation of 600 tonnes of carbon per km^2 and year.

[47] D.O. Hall, H.E. Mynick and R.H. Williams, 'Cooling the Greenhouse with Bioenergy', *Nature*, Vol. 353, pp. 11–12, 1991.

[48] K. Blok, W.C. Turkenburg, C.A. Hendriks and M. Steinberg, *Proceedings of the First International Conference on Carbon Dioxide Removal*, Amsterdam, 1992.

[49] S.C. Smelser, R.M. Stock and G.J. McCleary, *Engineering and Economic Evaluation of CO_2 Removal from Fossil Fuel Fired Power Plants*, Electric Power Research Institute, Report No. IE-7365, Vol. 1, Palo Alto CA, 1991.

[50] W. Fulkerson, R.R. Judkins and M.K. Sanghvi, 'Energy from fossil fuels', *Scientific American*, Vol. 263, No. 3, 1990, pp. 128–135.

and oil, would reduce emissions further. Advanced generating technology with natural gas as fuel can reduce emissions to about 0.10kgC/kWh, with additional benefits in lower emissions of other pollutants.

The main non-fossil alternatives for the provision of energy services include nuclear, wind, hydro, solar and biomass based power, and higher efficiency in the conversion, transport and end use of energy carriers. See Section 7.

6. Implications for the energy industries

Current efforts to reduce emissions of greenhouse gases, acid deposition precursors and photo-oxidant precursors are inadequate to achieve the emissions reductions needed to reach environmental goals formulated by governments. These goals include, for example, urban air quality guidelines, critical load levels and targets for reductions in NO_x and greenhouse gas emissions. The goals and demands for reducing emissions are likely to become more stringent as more is learned about the health and environmental impacts, and as it is increasingly recognized that policy decisions must be made also in situations where scientific knowledge is incomplete.[51]

Environmental goals and efforts to reduce emissions are based not only on current knowledge about sources and effects, but also on the perceived costs of emissions reductions; on the perceived costs of the environmental damage, exemplified by recent efforts to quantify external costs;[52] and on the level of protection for which societies are willing to pay. The limitations of cost-benefit analyses and the system boundaries (i.e. what costs and what benefits are accounted for) must therefore be carefully specified.

Current Long-Range Transboundary Air Pollution (LRTAP) protocols are aimed at single pollutant species and take into account acidification effects and ground level ozone formation. The Framework Convention on Climate Change addresses all greenhouse gases but does not commit the parties to the

---

[51] J. Ravetz, 'Usable knowledge, usable ignorance: incomplete science with policy implications', in W.C. Clark and R.E. Munn (eds.), *Sustainable Development of the Biosphere*, Cambridge University Press, Cambridge, 1986.
[52] O. Hohmeyer, *Social Costs of Energy Consumption*, Springer Verlag, Berlin and Heidelberg, 1989; R.L. Ottinger, D.R. Wooley, N.A. Robinson, D.R. Hodas and S.E. Babb, *Environmental Costs of Electricity*, prepared by Pace University Center for Environmental Legal Studies, Oceana Publications, New York, 1990.

Convention to any substantial emissions reductions. If, in cost-benefit analysis, particular pollutants are considered in isolation, the impacts of emission control on the groups of pollutants and effects are not evaluated and credit not given. For example, eutrophication is also reduced when reducing NO_x and SO_x emissions to reduce acidification. A reduction in European NO_x emissions reduces low-level ozone. A reduction in CO_2 emissions reduces SO_2 emissions due to the corresponding shift away from, notably, coal and oil.[53]

Many countries have proposed that the LRTAP protocols should be based on a combined emissions reductions approach. This is appropriate given the combined effects of pollutants, and the combined benefits of reductions. A combined emissions reductions approach would favour energy efficiency and a shift to natural gas and renewable sources of fuels and electricity. These are strategies that can simultaneously reduce emissions of greenhouse gases and the gases covered by the different LRTAP protocols. It should be also taken into account that there would be further improvements in many other fossil fuel related environmental problems, and in development and security considerations.

7. Options for new energy systems

The objective of an energy system is to provide energy services, at affordable cost and with socially acceptable side effects. Placing emphasis on energy services would bring into focus improved energy efficiency, especially at the point of end use.

Increased energy efficiency, particularly in the end use of energy, is a key strategy to meet the demand for energy services in a cost-effective way.[54] A major effort by the biggest Swedish utility, Vattenfall AB, to assess the potential

---

[53] O. Rentz, H.D. Haasis, A. Jattke, P. Russ, M. Wietschel and M. Amann, *Impacts of Energy Supply Structure on National Emission Reduction Potentials and Emission Reduction Costs*, Institute for Industrial Production, University of Karlsruhe, UFOPLAN Reference No. 104 04 006, Karlsruhe, 1992.

[54] See, for example, T.B. Johansson, B. Bodlund and R.H. Williams (eds.), *Electricity*, Lund University Press, Lund, Sweden, 1989; J. Goldemberg, T.B. Johansson, A.K.N. Reddy and R.H. Williams, *Energy for a Sustainable World*, Wiley Eastern Ltd., New Delhi, 1988; L.J. Nilsson, *Energy Systems in Transition*, PhD thesis, Department of Environmental and Energy Systems Studies, Lund University, Lund, Sweden, 1993.

for electricity conservation in Sweden concluded that 10 to 15 per cent could be saved through cost-effective overnight retrofit measures.[55] The same study estimated that lighting and air-handling energy use could be reduced by at least 50 per cent, without economic penalties, if energy-efficient equipment was introduced at the rate of capital turnover and expansion. Results from a project by the Californian utility Pacific Gas & Electric indicate that savings of as much as 75 per cent are technically achievable at costs competitive with energy supply, if appropriately designed efficiency packages are used.[56] Energy-efficient houses that use only about $25kWh/m^2/yr$ for space heating, compared to about $100kWh/m^2/yr$ for the stock average, can be built for Swedish climatic conditions without increasing the building's life-cycle cost or having unpleasing visual effects. Energy efficiency improvements are a very important and effective component in any least-cost no-regrets strategy to reduce CO_2 and other fossil fuel related emissions.[57]

Attention must, of course, be paid also to the potential health and environmental impacts from energy efficiency improvements. These include releases of mercury from inadequate handling of fluorescent tubes, and the various CFC and hydrochlorofluorocarbon (HCFC) compounds used in refrigerators and heat-pumps. An overriding concern at present is poor indoor air quality, where tighter building codes have been held responsible. However, it must be noted that good indoor air quality is not in conflict with energy conservation if low-emission building materials are used and adequate ventilation, with heat recovery if appropriate, is provided for.

On the supply side, nuclear energy has often been advanced as a major option. However, the expanded use of nuclear fission would require that a new generation of safer reactors is developed, that radioactive wastes can demonstrably be managed safely, and that weapons usable nuclear materials can effectively be safeguarded against diversions. Reactor technologies and fuel cycles that satisfy the first two criteria may be possible. Satisfying the third criterion may require

[55] A. Göransson, U. Lindahl, G. Forsman and C. Hedenström, *Commercial Buildings and Energy Conservation*, Report from the statistical energy audit study, Vattenfall AB, Sweden, 1991 (in Swedish).
[56] See Chapter 12 by Carl Weinberg.
[57] E. Mills, D. Wilson and T.B. Johansson, 'Getting started: No-regrets strategies for reducing greenhouse gas emissions', *Energy Policy*, Vol. 19, No. 6, pp. 526–42, 1991.

major technology and institutional changes, including international ownership and control of large parts of the nuclear enterprise.[58]

Fusion energy systems would be easier to safeguard against diversions of weapons useable material.[59] However, the authors of recent evaluations of fusion programmes agree that 40 to 50 years will be needed before fusion technology reaches commercialization.[60]

The other major supply option is renewable sources of energy. It has been suggested that given adequate support, these sources can contribute 40 and 60 per cent of global demand, in an energy-efficient world, for fuels and electricity respectively in the year 2050 at costs lower than those usually forecast for conventional sources of energy.[61] By that time, much of the present energy system will be replaced anyway.

Large hydro resources remain in Asia, South America and Africa. Spread of disease, population resettlement and effects on climate and on flora and fauna are problems that need increased attention when exploiting hydro resources.

Promising uses of biomass for energy include combined heat and power production, methanol production through gasification, and ethanol production through enzymatic hydrolysis. One potentially important source of biomass is agricultural and forest industry residues. The main environmental considerations associated with dedicated energy plantations are land-use requirements, use of fertilizers and recycling of ashes to maintain soil fertility, use of pesticides and herbicides, and atmospheric emissions during conversion. Potential environmental problems can be avoided with properly managed plantations and modern conversion technology.[62] In the case of deforested or

[58] R.H. Williams and H.A. Feiveson, 'Diversion-resistance criteria for future nuclear power', *Energy Policy*, Vol. 18, No. 6, pp. 543–49, 1990.

[59] J. Holdren, 'Safety and Environmental Aspects of Fusion Energy', *Annual Review of Energy and the Environment*, Vol. 16, 1991, pp. 235–58.

[60] U. Colombo and U. Farinelli, 'Progress in Fusion Energy', *Annual Review of Energy and the Environment*, Vol. 17, 1992, pp. 123-59.

[61] T.B. Johansson, H. Kelly, A.K.N. Reddy and R.H. Williams (eds.), *Renewable Energy, Sources for Fuels and Electricity*, Island Press, Washington DC, 1993.

[62] J. Beyea, J. Cook, D.O. Hall, R.H. Socolow and R.H. Williams, *Toward Ecological Guidelines for Large Scale Biomass Energy Development*, report of a workshop for engineers, ecologists and policymakers convened by the National Audubon Society and Princeton University, 6 May 1991, National Audubon Society, New York, 1991.

otherwise degraded lands, biomass plantations can make substantial ecological improvements.

A large number of different technologies and applications are available for direct and indirect (through thermodynamic cycles) uses of solar radiation. Solar-thermal electricity production is already almost commercially viable in areas with good clear-sky conditions. Photovoltaic technology is highly innovative and under rapid development. The principal environmental effect in both cases is surface area requirements.

Wind power is already economically competitive with conventional options for power production in some conditions, and costs are projected to continue to decrease through technical improvements and institutional learning. The environmental problems include noise, bird strikes, electromagnetic interference, safety, and the aesthetic considerations of visual impact. In most cases, environmental impacts have proved to be minor or negligible, or can be avoided through careful siting. The visual and audio impacts, which are a matter of personal evaluation, appear to be the most serious problems.

The relative lack of attention given to renewable sources of energy and energy efficiency in the past also means that there is much room for technology development. The lack of attention is indicated by IEA government spending on energy research where only 6 per cent of total research, design and development expenditures was allocated to renewables and 7 per cent to energy conservation in 1990.[63]

8. Concluding remarks

Substantial reductions are required in emissions causing urban air pollution, acidification and climate change in order to meet set objectives. This analysis suggests that end-of-pipe technical fixes are insufficient. Money spent today on developing less polluting coal-fired power plants, or tailpipe clean-up systems for petroleum driven internal combustion engine vehicles will not make sufficient contributions to solving the environmental problems in the longer run. It appears that action needs to be system wide with emphasis on

[63] International Energy Agency, *Energy Policies of the IEA Countries, 1991 Review*, IEA, Paris, 1992.

energy end use efficiency improvements and increased use of renewable sources of energy.

National action programmes, with close international collaboration, are required. Improvement of energy efficiency and use of modern conversion technology for renewables are particularly important strategies in developing countries where the energy sector is growing rapidly, in order to avoid building energy system infrastructures with the same deficiencies as those of the industrialized countries. Industrialized countries are more likely to introduce new technology successively when existing systems are replaced as they reach the end of their technical lifespan.

The rate at which new generations of advanced and efficient end-use and renewables technologies are commercialized would be enhanced if national and supra-national policies were redirected to reflect the important role that these technologies must play in the future energy system. The health and environmental impacts resulting from an increased use of renewables and greater emphasis on energy efficiency will, if properly managed, be minor or negligible. Pursuing such strategies would not only alleviate the range of environmental problems resulting from fossil fuel use, but would also improve the situation with respect to development and security issues.

Acknowledgements

In addition to the workshop participants, we thank Dean Abrahamson and Amulya Reddy for useful comments and constructive critique on earlier versions of this paper. We are also grateful to Mark DeLuchi for providing data for Figure 6.

Economic Impacts of Environmental Law: The US Experience and its International Relevance

Valerie M. Fogleman

Abstract

Since 1970 the US energy industry has been increasingly constrained by environmental statutes and regulations. In the late 1960s there were few environmental constraints on the industry, but most aspects of it are now highly regulated by an ever-increasing array of environmental laws.

The economic cost to the energy industry of complying with the environmental requirements of the past 23 years has been huge, and indications are that the cost will remain high for the foreseeable future. For example, despite the strict regulation of air and water pollution, recent studies indicate that 54 million people in the US still live in areas in which the air harms their health,[1] and half of the country's streams are still not suitable for fishing or swimming.[2] Another recent study has estimated that the US oil industry must spend over $166 billion between 1991 and 2010 merely to comply with existing and anticipated environmental requirements.[3]

This paper briefly examines the major environmental laws which have affected the oil, gas, coal, hydroelectric power, and nuclear energy industries in the US since 1970 as well as discussing proposals for further environmental

---

[1] *Environment Reporter Current Developments*, Bureau of National Affairs, Washington DC, 12 November 1993.

[2] *Environment Reporter Current Developments*, Bureau of National Affairs, Washington DC, 29 October 1993.

[3] National Petroleum Council, *US Petroleum Refining: Meeting Requirements for Cleaner Fuels and Refineries*, NPC, Washington DC, August 1993.

laws. It also investigates the economic impact which these laws have had on the US energy industry.

Obviously, the experience with environmental constraints which the energy industry has had in the US will not be mirrored elsewhere because the structure of the industry differs in other countries. However, as environmental constraints on the energy industry increase worldwide, it is useful to review the US experience, not least in an attempt to avoid many of the expensive confrontations which have accompanied it.

Introduction

Since the 1970s the energy industry in the United States has been required to comply with increasingly complex and strict environmental laws. Whereas these laws reflect an ever-growing environmental awareness on the part of the US Congress and the American people, they also reflect a nation struggling to balance the goals of economic growth and environmental quality.

This paper will briefly examine the goals and evolution of the major environmental laws which govern the way in which the US energy industry obtains required permits from the federal government; emits pollutants into the air; discharges effluent into waterways; disposes of hazardous waste; and accounts for hazardous waste disposed of in the past. In other words, the paper will examine the effects of US environmental laws on the operation of the energy industry in the United States. It will conclude by discussing the international relevance of the US energy industry's experience in complying with federal environmental laws.

Federal permitting

The energy industry in the United States arguably first became aware of the potentially significant effect of federal environmental statutes in 1971 when the District of Columbia Circuit Court of Appeals, in *Calvert Cliffs' Coordinating Committee, Inc. v. Atomic Energy Commission*, held that the Atomic Energy Commission must comply with the National Environmental Policy Act (NEPA) in considering applications for construction permits and operating licences for nuclear power plants.[4]

NEPA had become law in 1970,[5] and requires all federal agencies to consider the significant environmental effects of proposed major federal actions before deciding to conduct the actions. Thus, for example, the appropriate federal agency must comply with NEPA prior to granting a licence for a hydroelectric project; granting a lease to drill an oil or gas well on federal land; approving the right of way for an electricity transmission line over federal land; or leasing

---

[4] Vol. 449, Federal Reporter Second Series, District of Columbia Circuit Court of Appeals, 1971, p.1109.
[5] Public Law No. 91-190, Vol. 83, Statutes at Large, 1970, p.852 (codified at title 42 United States Code, sections 4321-4347)

a section of the Outer Continental Shelf for oil exploration. Compliance entails considering whether the proposed action may have any significant environmental effects and, if so, preparing an environmental impact statement (EIS). The EIS may take months or years to prepare and requires extensive public input. The requirements of what it should cover are defined by regulations issued by the Council on Environmental Quality.[6]

Whereas NEPA serves an admirable purpose in ensuring that federal agencies consider the environmental impacts of their actions, it provides opponents of a proposed energy (or other) project with the means to delay the project, often for many years, by challenging the agency's compliance with NEPA's procedural requirements. Even if the opponents ultimately lose the lawsuit, they may have achieved their objective because the project may be abandoned for reasons of cost. This strategy, which has become known as 'winning by losing slowly', has been used to great effect and has cost the energy industry millions, if not billions, of dollars.

The abandonment of the proposed SOHIO pipeline which was to have transported an estimated 500,000 barrels of crude oil per day from Alaska to Midland, Texas, is an example of the cost to the energy industry of environmental legislation (although this was not exclusively due to NEPA). In the 1970s SOHIO spent over $50 million and four years attempting to obtain 700 federal, state and local permits which it needed to construct and operate the pipeline. SOHIO subsequently abandoned the pipeline in 1979, however, mainly because of delays caused by the need to comply with environmental requirements.[7]

Emission of air pollutants

Although individual energy industries may be affected more by statutes concerning, for example, the reclamation of mining sites, as a general rule the environmental statutes which have had the most impact on the US energy

---

[6] Vol. 40, Code of Federal Regulations, part 1500. The Clinton Administration's proposal to abolish the Council on Environmental Quality, even if enacted by Congress, would not have any impact on the continued effect of the Council's regulations.

[7] See generally G. Lew and L. Silverstrom, 'Energy and Environmental Law in Transition: A Practitioner's Perspective', *Energy and Environmental Law 1980*, p. 818.

industry are the Clean Air Acts.[8] This is because the production and consumption of energy pollutes the air more than any other environmental medium.

The Clean Air Act Amendments of 1970, therefore, had a major economic effect on the US energy industry. This legislation revolutionized the manner in which air pollution was regulated in the United States. Before these amendments, the statutes which purported to regulate air pollution in the US had proven to be ineffective, with the result that air quality across the country was deteriorating.

Congress decided that the way to counteract the deterioration in air quality was to establish a 'command and control' system to be implemented by a newly created federal agency known as the Environmental Protection Agency (EPA) in cooperation with the states.

The 1970 Amendments authorize the EPA to establish national ambient air quality standards (NAAQS) for pollutants such as sulphur dioxide and carbon monoxide and also to establish standards for hazardous air pollutants, as designated by the EPA. Under the 1970 Amendments, each state prepared a plan setting out the means by which it would implement the legislation to achieve NAAQS within that state within a certain period of time. The state programmes were required to be at least as stringent as the federal one. If the EPA approved a state's plan, the state administered and enforced it; otherwise, the EPA was to administer and enforce the federal programme.

The Act required facilities which emitted air pollutants to be licensed, with the licences specifying the type and level of pollutants permitted from each emission source at the facility.

The 1970 Amendments also established a regulatory system for emissions from 'moving sources' such as motor vehicles and aircraft. This system was implemented by the EPA and was designed, in part, to require the reduction of emissions of pollutants by means of phased-in design requirements for motor vehicles, motor vehicle engines, fuel, and fuel additives.

The 1970 Amendments provided for civil and/or criminal penalties for non-compliance with the Act. Additionally, they contained 'citizen suit' provisions; that is, they authorized private persons or groups such as environmental organizations to sue a regulated facility for non-compliance with the Act if

---

[8] Title 42, United States Code, sections 7401 *et seq*.

the EPA failed to require compliance. The 1970 Amendments also authorized private persons to sue the EPA to require it to conduct an action mandated by the Act, such as promulgating a regulation, if the EPA failed to perform the action. Thus, the 1970 Amendments led to the creation of what are sometimes known as 'private attorneys general' by authorizing lawsuits by what have become powerful and influential environmental organizations. These organizations are funded by various means including donations from charitable organizations, individuals and others. Additionally, lawyers and other persons may donate their time to specific lawsuits and, since the passage of the Equal Access to Justice Act of 1980,[9] attorney fees may be awarded in successful cases against the US government.

The 1970 Amendments were designed to be 'action or technology forcing', that is, they required industries regulated by them to install and continually update pollution control technology to reduce the emission of air pollutants. Congress obviously realized the tremendous cost which it was requiring industry to bear under threat of the closure of individual plants. However, any concern regarding the cost involved was apparently secondary to the determination of Congress to reduce air pollution. Indeed, the statutory criteria which the EPA must use to establish NAAQS are based on considerations of human health and welfare, not economics or technology.

The obvious result of legislatively imposing huge costs on American industry was litigation. For example, the electric utilities, which were particularly affected by the controls on emissions of sulphur dioxide, judicially challenged the EPA's interpretation of the 1970 Amendments in EPA regulations.[10] The EPA had felt that in approving state implementation plans it could not consider economic or technological feasibility of compliance with the NAAQS. The implementation of the 1970 Amendments became marked, therefore, not by cooperation between regulators and the energy and other regulated industries but by expensive and protracted conflicts.

Perhaps not surprisingly, the goals of the 1970 Amendments were not achieved within the designated time frame. By the mid-1970s, the EPA was encountering serious enforcement problems. In 1977, therefore, when Congress

[9] Title 28, United States Code, section 2412(d).
[10] Union Electric Company versus Environmental Protection Agency, United States Reports, Vol. 427, 1976, p. 246.

amended the Clean Air Act again, it had no option but to establish 'non-attainment' areas, that is, areas which had failed to attain designated standards within the designated time period and which were given new deadlines. In turn, creation of non-attainment areas meant expanding the command and control system to accommodate the added complexity of the legislation.

The 1977 Amendments also contained a market-based concept, albeit to a limited degree. Under the concept, which has become known as 'emissions trading', a new pollution source which is proposed for a non-attainment area must obtain from existing pollution sources in that area an agreement that they will reduce their pollution by a greater amount than the new source will emit. The 1977 Amendments, therefore, encouraged new sources to trade the right to emit a certain amount of designated pollutants with existing sources. Although the market created for emissions trading was severely limited because of the narrow scope of parties which were able to participate, it marked the beginning of a partial market-based alternative to the complex and expensive command and control system. As indicated earlier, all stationary sources which emit air pollutants must be licensed, whether or not they are constructed in non-attainment areas.

In 1990, after prolonged debate, Congress again amended the Clean Air Act, more than doubling the size of the legislation and requiring the EPA to promulgate (and the regulated community to comply with) a daunting number of new regulations over the next several years. One estimate of the *administrative cost* of complying with the federal permit programme of the 1990 Amendments comes to $500 million per year.[11] (The federal permit programme is designed to enhance the EPA's ability to oversee all Clean Air Act programmes.)

Unfortunately, Congress has not been deterred from imposing rigid deadlines regarding the time by which the EPA must promulgate regulations and the time by which the regulated industries must comply. The 1990 Amendments are filled with strict deadlines. By May 1992, environmental organizations had already notified the EPA of their intention to file suit over 50 regulations which had allegedly been delayed past their congressionally mandated deadlines.

---

[11] *Environment Reporter Current Developments*, Bureau of National Affairs, Washington DC, 29 May 1992.

The 1990 Amendments significantly expand the command and control system of earlier Clean Air Acts, both in the number of pollutants regulated by the EPA and the scope of activities to which the Act applies. For example, the 1990 Amendments designate 189 substances as hazardous air pollutants in contrast to the eight designated by the EPA under the earlier Acts. In what has come to be an inevitable consequence of the amendment of major environmental statutes, the 1990 Amendments increased the scope and scale of liabilities for non-compliance.

Significantly, however, the 1990 Amendments have expanded the use of emissions trading introduced in the 1977 Amendments. In a major effort to reduce acid rain, the 1990 Amendments established a system whereby electric power plants which are required by the 1990 Amendments to reduce their emissions of sulphur dioxide by 50 per cent, receive pollution credits for the amount of emissions by which they exceed the reduction. The pollution credits are tradeable in units comprising allowances of one tonne of sulphur dioxide, and may be sold to operators who do not anticipate being able to achieve the 50 per cent reduction or who would prefer time to meet the reduction additional to that which would otherwise be imposed by the 1990 Amendments. Such operators are required to purchase sufficient permits to offset their emissions to the extent by which these exceed the allowances specified in the 1990 Amendments and their licences, in order to avoid penalties. The EPA was also authorized to create a set number of pollution credits to sell to utilities which require additional time to comply with the 1990 Act.

The first auction of pollution rights was held in March 1993 by the Chicago Board of Trade on behalf of the EPA. At the auction, a total of $21.4 million in pollution rights was purchased, nearly all being EPA-offered allowances which had no minimum bid level, as opposed to privately offered allowances which did have such a level. The purchasers of the allowances bought the rights to emit 150,010 tons of sulphur dioxide after 1995.[12]

The expansion of emissions trading in the 1990 Amendments is a significant departure from the command and control scheme of other provisions of the Clean Air Act. Emissions trading will not replace the command and control

---

[12] *Environment Reporter Current Developments*, Bureau of National Affairs, Washington DC, 2 April 1993.

scheme, not only because that system is so well established, but because there would be a reluctance to purchase pollution rights if a purchaser knew that the regulatory authority did not intend to enforce the licensing system under which a facility is permitted to emit only a certain amount of pollutants. The emissions trading system has the potential, however, to result in a reduction of air pollution at lower cost to the energy industry than implementation of the command and control system. If the system is extended, as is happening in states such as California, Massachusetts, New Jersey and Washington,[13] the system has the potential to permit credits to be made available by those industries which can reduce emissions for the least cost and to be purchased by those which find emission control more expensive to achieve.

In another development, the cost of complying with the Clean Air Act may also be reduced by an EPA initiative to encourage voluntary cooperation by the regulated industries in place of additional rule-making. Further, the EPA is increasingly engaging in what has become known as 'regulatory negotiation' or 'reg neg' for certain of its rule-making procedures. Regulatory negotiation is the active participation in formulating regulations by persons with an interest in them, such as industry, states, and environmental organizations. A major advantage of regulatory negotiation for the regulated industry is the promulgation of more considered regulations, whereas a major advantage for the EPA is the reduced risk of judicial challenge to its regulations.

In summary, air pollution in the United States is subject to a highly complex command and control system. Although this system has resulted in a significant reduction in air pollution, it has done so at a tremendous cost to the US energy and other industries. While the command and control system is unlikely to be changed, there is an increasing tendency for the US Congress and the EPA to attempt to cooperate with the regulated industries in certain aspects of it. Examples of this cooperation are the introduction of emissions trading, voluntary agreements between the EPA and the major regulated industries in place of rule-making, and the introduction of regulatory negotiation. Extended use of these concepts will – it is hoped – reduce the tremendous cost of complying with the ever-tightening emission controls.

---

[13] *Environment Reporter Current Developments*, Bureau of National Affairs, Washington DC, 1 October 1993.

Discharges of effluent

Just as the energy industry and other industries in the United States must apply for permits to emit pollutants into the air, they must apply for permits to discharge effluent from 'point sources', that is from ditches or pipes into navigable waters. The legislation establishing the permit system, which is known as the national pollutant discharge elimination system (NPDES), has many similarities to the emissions control system of the Clean Air Act, both in its structure and in its economic effect on the industries regulated by it.

The NPDES programme was introduced in the Federal Water Pollution Control Act Amendments of 1972.[14] These focused on the EPA as the central actor in the system, while inducing states to develop state programmes by, for instance, authorizing those which had an EPA-approved NPDES programme to administer and enforce their programmes in place of the federal programme.

Other provisions of the 1972 Amendments established a programme whereby the Corps of Engineers issues permits for dredge and fill activities in navigable waters; authorized a federal funding programme, implemented by the EPA, for the construction of publicly owned treatment works; and created civil and/or criminal penalties for non-compliance with the Act. Citizen suit provisions were also included, being modelled almost verbatim on those in the Clean Air Act Amendments of 1970.

Like their Clean Air Act counterpart, the 1972 Amendments were 'action or technology forcing', requiring industry to install continually updated pollution control equipment to reduce the discharge of pollutants into waterways. The method adopted by the 1972 Amendments was to require regulated industries to limit the discharge of pollutants by means of the 'best practicable control technology currently available' or other such standards depending on the pollutant involved. These standards were to be continually tightened to force the continued reduction of discharges of pollutants.

Like those of the Clean Air Act Amendments of 1970, the goals of the Federal Water Pollution Control Act Amendments of 1972 were over ambitious and proved impossible to achieve in the period envisaged by Congress. Indeed, the 1972 Amendments' unrealistic goal of eliminating the discharge of pollutants into navigable waters by 1985 indicates an alarming naïveté on the

---

[14] Title 33, United States Code, sections 1251 *et seq.*

part of Congress in its understanding of the economic burden it was imposing on the regulated industries.

The expensive and protracted litigation which accompanied the Clean Air Act Amendments of 1970 also characterized the implementation of the 1972 Amendments, a result which was inevitable due to the huge costs imposed by the complex and over-ambitious legislation.

In 1977 when Congress amended the 1972 Amendments by enacting the Clean Water Act it realized that it had no choice but to extend the date by which certain water quality standards were to be met. Unfortunately, this realization did not occur before the energy and other industries had spent huge sums in attempting to comply with the standards established by the unrealistic deadlines of the 1972 Amendments.

The 1977 Act extended the complex command and control system. Among other things, it created a new classification of pollutants and a new effluent limitation standard and established a timetable for continually revising the standard. Thus the strategy begun by Congress in the 1972 Amendments was continued and extended at a continually increasing cost to the American energy and other industries.

In 1987 Congress again reauthorized and amended the Clean Water Act resulting in an increase of its scope and more severe penalties for noncompliance.

Hearings have begun in preparation for a further reauthorization of the Clean Water Act. One proposed provision which is proving to be controversial is the effluent guidelines provision, under which the EPA would have authority to intervene in the design and operations of a facility in order to require the reduction of hazardous effluent by a change in the production process, product or raw materials. Industry has also raised concerns about the proposed provision which could empower the EPA to identify highly toxic bioaccumulative pollutants and ban their discharge.

There is a possibility, however, that some provisions in the Clean Water Act reauthorization bill may introduce market-based concepts. The EPA has indicated that the introduction of such concepts could be feasible, for example, in controlling discharges of nutrients on the basis of the watershed in which those nutrients are discharged. Additionally, the EPA has stated that it will pursue a partnership approach rather than relying solely on the command and control system of regulating water pollution.

Disposal of hazardous waste

In 1976, Congress expanded, modified, and reorganized earlier legislation on waste in a manner which was similar to its earlier legislation regulating air and water pollution. The Resource Conservation and Recovery Act of 1976 (RCRA) established cradle-to-grave controls for hazardous waste by regulating the material from its generation to its ultimate disposal.[15]

To implement and enforce the hazardous waste controls, RCRA created an Office of Solid Waste within the EPA, enlarging the agency and reinforcing its central role in implementing most federal environmental legislation in the United States.

RCRA has many conceptual similarities to the legislation on air and water pollution. The generation, handling, treatment and disposal of hazardous waste is regulated by a complex command and control system. Among other requirements of RCRA, the EPA sets standards for the treatment and disposal of hazardous waste and identifies classes of persons, such as generators and transporters of hazardous waste and operators of treatment, storage and disposal facilities, who must comply with the provisions of RCRA under threat of civil and/or criminal penalties.

As with the clean air and clean water legislation, states administer and enforce programmes regulating hazardous waste within their boundaries upon the EPA's approval of their programmes. Additionally, the citizen suit provisions of RCRA are similar to those in the clean air and clean water legislation.

Also like its counterparts which regulate air and water pollution, RCRA has been amended extensively. In 1980, Congress enacted the Solid Waste Disposal Act Amendments which expanded the penalties for failing to handle hazardous waste according to federal requirements. To aid in enforcing the requirements, the 1980 Act amended the search and seizure provisions to facilitate the scope of its investigations. Indeed, the increase in the scope and scale of civil and criminal liability has become almost inevitable each time an environmental statute is amended by Congress.

In 1984, Congress again extensively amended RCRA with the enactment of the Hazardous and Solid Waste Amendments of that year. These imposed new requirements for the land disposal of hazardous waste, obliging the EPA

---

[15] Title 42, United States Code, sections 6901 *et seq.*

to issue regulations regarding those requirements on a strict timetable; they also extended the scope of RCRA to cover small quantity generators of hazardous waste, and amended the requirements for treatment, storage and disposal facilities.

The cost to the energy and other industries of complying with RCRA is enormous. For example, the corrective action programme which was established by the Hazardous and Solid Waste Amendments of 1984 will, when implemented, require owners or operators of treatment, storage and disposal facilities to clean up at the sites. These provisions do not affect only a discrete industry which owns or operates treatment, storage and disposal facilities for use by third parties because industries which generate hazardous waste may own or operate a disposal site specifically for hazardous waste generated by them. Estimates of the cost of the corrective action programme range from $7 billion to $850 billion.

The high cost of compliance is not the only similarity between RCRA and the air and water pollution legislation. Like those other statutes, RCRA contains provisions which have proven unrealistic. The clean-up provisions and provisions ordering the EPA to promulgate regulations to require owners or operators of underground storage tanks for petroleum to obtain financial security (that is, insurance or other means of evidencing financial responsibility) for clean-up and third-party liability, and to comply with RCRA's technical standards for tanks, failed to consider adequately the prohibitive cost of compliance encountered by many owners or operators of small petrol stations.

As a result of the problems encountered by this latter group, the EPA has been required repeatedly to delay implementation of the financial responsibility requirements of RCRA. In order to prevent the closure of large numbers of such businesses, the states have attempted to establish programmes to grant financial aid to owners or operators of small petrol stations in cleaning up leaked petroleum products. Some of these funds have, however, been overwhelmed by the huge costs involved.

The underground storage tank provisions of RCRA illustrate another of its similarities to the air and water pollution legislation, that is, a requirement for the regulated industries to update their equipment to minimize pollution. The 'land ban' requirements of RCRA are, however, perhaps the most costly aspect of this in terms of forcing technology. These requirements also illustrate the tension which exists between Congress and the EPA, and the influence of environmental organizations.

Basically, Congress required the EPA to issue regulations to ban the land disposal of hazardous waste by certain congressionally imposed deadlines. The regulations which the EPA has promulgated in this case are complex. They establish among other things, the treatment which hazardous waste must undergo before its disposal. Under the influence of environmental organizations which cited evidence of the EPA's failure to comply with the deadlines of other environmental statutes, Congress banned the land disposal of hazardous waste after the deadlines unless the EPA had promulgated the regulations by those dates. The economic cost of the land ban regulations is huge and, in some cases, has tripled the cost of hazardous waste disposal.

It appears probable that the cost to industry of complying with RCRA will increase substantially after its next reauthorization. In July 1992, the EPA estimated that a reauthorization bill approved by a Senate Committee would cost consumers between $8 billion and $46 billion per year, excluding the cost to governmental agencies of implementing it. The EPA noted that the bill provided little risk reduction and did not target significant risks to human health and the environment.

Accounting for hazardous waste disposed of in the past

The enactment of RCRA and its amendments did not address hazardous waste which had been disposed of to less exacting standards before the Act came into force. In 1980 Congress focused on the problems for human health and welfare and for the environment caused by historic pollution, by enacting the Comprehensive Environmental Response, Compensation and Liability Act (CERCLA).[16] This provides the EPA with the power and the funds to remove or deal with hazardous waste and to require persons connected with the waste to pay for its removal or treatment.

Basically, CERCLA authorizes the EPA to use monies from a 'Superfund' to clean up hazardous substances and to require certain classes of persons known as potentially responsible parties (PRPs) to reimburse the EPA for the

---

[16] Title 42, United States Code, sections 9601 *et seq*.

cost of cleaning up the hazardous substances. As an alternative, CERCLA authorizes the EPA to require PRPs to clean up the hazardous substances. Liability under CERCLA is strict, retroactive and, unless a PRP can prove that there is a reasonable basis for apportioning clean-up costs, joint and several; that is, a PRP may be liable for the entire cost of cleaning up a site in which it is involved, although it has a right to sue other PRPs – which it believes to have a responsibility – for a portion of these costs. The liability system is so broad, in fact, that it has exceeded traditional liability systems to become a quasi-taxation system, imposing a retroactive 'tax' on anyone associated with sites which are cleaned up under CERCLA.

There are four categories of PRPs: current owners or operators of contaminated sites; owners or operators of contaminated sites at the time at which there was a disposal of hazardous substances; persons who 'arranged for' the disposal of hazardous substances; and persons who transported hazardous substances to a site selected by them.

As with all other environmental legislation in the United States, the energy industry is particularly affected because the production of energy necessarily involves the disposal of large amounts of hazardous substances. For example, electric utilities are being required to pay for the clean-up of sites at which dealers who purchased used transformers from the utilities stripped the transformers for copper, in the process spilling fluid containing polychlorinated biphenyls. Mining sites have been added to the EPA's list of the most contaminated locations despite the relatively low concentrations of hazardous substances which are present in mining waste.

The cost to the energy and other industries of cleaning up past pollution is vast. For example, the average cost of cleaning up a site on the EPA's list of the most contaminated sites is between $30 million and $50 million. One estimate of the total cost of cleaning up historic hazardous waste in the US exceeds one trillion dollars, and much of this may have to be borne ultimately by the insurance and reinsurance industry unless a congressional solution is enacted.

CERCLA does not, for the most part, create a command and control system because the majority of its provisions concern primarily the clean-up of hazardous substances disposed of in the past rather than the regulation of any ongoing activities. CERCLA does, however, reflect some of the faults of other environmental statutes.

When Congress enacted CERCLA, it did so under the misapprehension that there were relatively few hazardous waste sites in the United States. Accordingly, Congress established a $1.6 billion Superfund which has had to be increased and will have to be funded for many, many years. Congress did not merely fail to realize how many sites needed cleaning up, it also underestimated the length of time involved. As a result, the EPA has been under intense pressure to speed up the clean-up of sites.

The realization by Congress, when it amended CERCLA in 1986, that much more money and time was needed to clean up the United States' past hazardous waste problem did not prevent Congress from increasing the cost of the programme to the affected industries, however. The Superfund Amendments and Reauthorization Act of 1986 (SARA) incorporated into CERCLA some of the stringent standards of other environmental legislation when, according to the EPA, substantive and promulgated requirements of federal or state environmental statutes and regulations are 'applicable or relevant and appropriate'. For example, SARA requires remedial actions to attain Maximum Contaminant Level Goals (MCLGs) designated under the Safe Drinking Water Act when the goal, according to the EPA, is relevant and appropriate to the circumstances of the release of a hazardous substance. The MCLG for carcinogens is zero, a level which it is impossible to detect in water with the scientific knowledge and instrumentation available at present. The EPA regulations, therefore, do not consider MCLGs of zero to be appropriate for cleaning up surface and ground water at CERCLA sites.

The requirement for the clean-up of sites to attain MCLGs when the goal is relevant and appropriate, however, demonstrates seeming determination of Congress to impose exacting standards on American industry regardless of the cost involved and the technological infeasibility of attaining the standard. Only in the last two or three years has there been any real recognition by Congress and the EPA that American industry simply does not have sufficient money to clean-up all hazardous waste sites to the standards which Congress continues to impose. Whether Congress will alleviate the economic burden in the scheduled reauthorization of CERCLA remains to be seen.

Perhaps the most glaring error of Congress in enacting Superfund, however, was its failure to foresee the consequences of imposing the strict, retroactive, joint and several liability system on American industry. A recent study has estimated that 88 per cent of the costs of Superfund matters are accounted for by transaction costs. Most transaction costs are legal costs involved in attempting to force other PRPs and insurers to pay clean-up costs rather than actually cleaning up hazardous waste.[17] Furthermore, the total number of claims and insurance coverage lawsuits continues to escalate: at the beginning of 1991, 13 of the top 20 property and casualty insurers reported having approximately 50,000 claims and 2,000 lawsuits pending over pollution claims.[18] Although a solution may result from the pending reauthorization of CERCLA, the difficulties to be overcome in reaching such a solution are numerous and complex.

The international relevance of the US experience

The discussion in this paper has focused on the major federal environmental statutes which affect the manner in which the energy industry operates in the United States. It has not attempted to describe the effects of all federal environmental statutes, partly because the list is so long, as illustrated by the partial list of such statutes in the appendix, and partly because the discussion of the major statutes provides an overall view of the system of regulating pollution caused by the energy industry in the United States.

Although the structure of the energy industries in other countries differs from that in the United States, the US experience in dealing with pollution problems offers valuable lessons, at least for privately owned energy industries in other countries. This is because very few, if any, of the pollution problems addressed by US environmental laws are unique to the United States. For example, the European Union is as concerned about reducing air pollution caused by emissions of sulphur dioxide, nitric oxide, volatile organic compounds, and carbon monoxide as is the United States. Hazardous waste disposal is no less of a problem in other industrialized nations than it is in the United States. Indeed, some problems such as discharges of pollutants into rivers may be more acute in other countries because of the greater number of

---

[17] Jan Paul Acton et al., *Superfund and Transaction Costs: The Experiences of Insurers and Very Large Industrial Firms*, RAND/The Institute of Civil Justice, Santa Monica, CA, 1992.
[18] *Hazardous Waste: Pollution Claims Experience of Property/Casualty Insurers*, GAO/ RCED-91-59, United States General Accounting Office, Washington DC, 5 February 1991.

rivers which are international, and so on. Additionally, pollution problems are increasingly seen to be global, as demonstrated by the threat of climate change and deterioration of the stratospheric ozone layer.

Perhaps the greatest lesson from the US experience for legislators in other countries is not to enact overly ambitious legislation which places unrealistic demands on regulated industries. Both the Clean Air Act and the Clean Water Act suffer from this failing. Although implementation of the statutes has achieved significant reductions in air and water pollution, they necessarily placed the regulators and the regulated in adversarial positions which, in turn, has necessarily resulted in expensive and time-consuming conflicts.

The US experience also shows that legislators must consider the cost to the regulated industry before enacting environmental statutes – both the cost of an individual environmental statute and the cumulative cost of all environmental statutes. The US environmental legislation demonstrates that expensive and time-consuming conflicts are inevitable when the cost of complying with environmental legislation is not considered adequately.

Estimates of the cost to the energy industry of complying with environmental legislation are staggering. For example, a recent study prepared for the US Department of Energy by the National Petroleum Refinery Association estimated that the cost to the US oil industry of complying with existing and anticipated environmental regulations during the period from 1991 to 2010 would be $166 billion.[19] This figure is particularly remarkable when it is considered that US industry has already spent 20 years in restructuring and in upgrading pollution control equipment in order to comply with US environmental legislation.

The US experience also shows, unfortunately, that even when Congress believes that it knows the cost of implementing environmental legislation, it may be disastrously wrong as in the case of the Superfund legislation described above. Experience now shows, of course, that initial estimates by Congress of the problem bear no relation to current estimates. If Congress had realized the actual cost to American industry of the Superfund programme in 1980, it might not have enacted CERCLA in its present form. The US experience with Superfund serves as a valuable lesson, therefore, to the European

---

[19] NPC, *US Petroleum Refining*, op. cit.

Commission and other government authorities as they consider the problem of cleaning up historic pollution and the allied problem of the standard to which the environment should be cleaned. In this regard, other countries should consider carefully the 'polluter pays' principle which is a cornerstone of CERCLA. Although this principle, which has been adopted by the Council of the European Union, sounds reasonable, it is too simplistic because every member of society is a polluter to some degree. Too often, the 'polluter pays principle' has become the 'industrial polluter pays principle'.

Another major lesson is that legislators must realize the complexity involved in a command and control system and the inevitable tendency of such a system to expand. The US Congress and the EPA have now recognized the advantages of market-based initiatives and cooperation with the regulated community, but this recognition has only come after numerous expensive and lengthy conflicts. The enactment of the Energy Policy Act in 1992 also demonstrates a belated recognition by the US administration and Congress of the environmental advantages of energy efficiency.[20]

Finally, the US experience demonstrates that once power is accorded to private citizens or groups such as environmental organizations to act as private attorneys general by bringing lawsuits, the power of such organizations will expand. The major, and many smaller, environmental organizations in the United States perform a valuable public role. However, before granting significant powers to such organizations in other countries, legislators must understand the effect of the exercise of those powers.

---

[20] Public Law No. 102–486, *Statutes at Large*, Vol. 106, 1992, p. 2776.

Appendix: Partial list of US federal environmental statutes affecting the energy industry

Statute	*Year enacted*
Rivers and Harbors Act	1899
Federal Water Pollution Control Act	1948
Outer Continental Shelf Lands Act	1953
Air Pollution Control Act	1955
Federal Water Pollution Control Act Amendments	1956
Fish and Wildlife Act	1956
Clean Air Act	1963
Wilderness Act	1964
Water Quality Act	1965
Solid Waste Disposal Act	1965
Clean Water Restoration Act	1966
Air Quality Act	1967
Wild and Scenic Rivers Act	1968
National Environmental Policy Act	1970
Water Quality Improvement Act	1970
Resource Recovery Act	1970
Clean Air Act Amendments	1970
Noise Control Act	1972
Coastal Zone Management Act	1972
Federal Water Pollution Control Act Amendments	1972
Marine Protection, Research and Sanctuaries Act	1972
Marine Mammal Protection Act	1972
Endangered Species Act	1973
Safe Drinking Water Act	1974
Toxic Substances Control Act	1976
Coastal Zone Management Act Amendments	1976
Resource Conservation and Recovery Act	1976
Deepwater Port Act	1976
Clean Water Act	1977

Clean Air Act Amendments	1977
Federal Surface Mining Control and Reclamation Act	1977
Endangered Species Act Amendments	1978
National Oil Pollution Planning Act	1978
Outer Continental Shelf Lands Act Amendments	1978
Comprehensive Environmental Response, Compensation and Liability Act	1980
Fish and Wildlife Conservation Act	1980
Solid Waste Disposal Act Amendments	1980
Used Oil Recycling Act	1980
Endangered Species Act Amendments	1982
Nuclear Waste Policy Act	1982
Hazardous and Solid Waste Amendments	1984
Superfund Amendments and Reauthorization Act	1986
Safe Drinking Water Act Amendments	1986
Emergency Planning and Community Right-to-Know Act	1986
Clean Water Act Amendments	1987
Nuclear Waste Policy Act Amendments	1987
Pollution Prevention Act	1990
Oil Pollution Act	1990
Clean Air Act Amendments	1990
Pollution Prosecution Act	1990
Energy Policy Act	1992

The Impact of Environmental Policy on Trade

Vincent Cable

Abstract

Trade and environment linkages are becoming a major focus for trade negotiation, particularly in the GATT.

There are some areas of broad agreement. Sustainable development is, in general, more likely to be achieved in an open multilateral trading system which enables developing countries to diversify their exports and allows discipline to be imposed on agricultural and other subsidies. Global environmental agreements may have to be buttressed by trade provisions, and these will need safeguards to ensure that measures are not used in a discriminatory and protectionist manner. The major application of this principle will come with action taken to counter global warming; carbon taxes are market instruments which are compatible with an open trading system. Countries should be free to set product standards which reflect their own environmental preferences, provided this does not involve discrimination against overseas suppliers.

The big area of controversy is 'environmental dumping': the demand from some environmentalists and industrialists that they should be protected from competition from 'low cost' suppliers who observe lower process standards. The paper argues strongly against admitting trade restrictions on these grounds, both for practical reasons and in principle. In any event, the current of trade and investment flows generated by differences in environmental standards seems small.

1. 'Green trade': a new battleground?

Before the dust settles on the recently completed GATT Uruguay Round, the next round is already being designed. There are strong pressures to paint it green. Environment became a serious bone of contention in the US–Mexico NAFTA agreement; supplementary side letters to the treaty introduced the possibility of imposing sanctions if Mexico fails to enforce its own environmental legislation. Some non-governmental organizations (NGOs), who wanted more by way of sanctions, were not satisfied and opposed the treaty anyway. NGOs, and a vociferous section of the US Congress, now want GATT rules to incorporate environmental concerns more explicitly in the workings of the World Trade Organization which will succeed GATT under the terms of the Uruguay Round Agreement. At the same time, many who are concerned to preserve an open trading system are becoming deeply alarmed at the prospect that environmentalism could become an excuse for protectionism and for trade measures of a damaging kind. There is thus the possibility of a serious clash of interests and ideologies over trade and environment (though there are some areas of common ground, including the need to act upon global and cross-border environmental hazards).

The debate centres on the existence of national differences in environmental preferences and standards. For those who favour a free, open trading system, such differences – like differences in climate, productivity, wages and working standards – are inevitable in a world of sovereign states and provide a rationale for mutually beneficial trade (or investment flows). They consider that this trade, over time, and with rising living standards in poorer countries, narrows the differences between countries. Others, however, regard such differences in environmental standards as offensive and trade based upon them as unacceptable. There are two strands to this concern.

The first is the idea that some 'lifestyles' issues are so important that one country has a right to impose its preferences on another, using trade as a lever. (There are analogies here with human rights as a trade and aid issue.) A few years ago, a group of fishermen from Mexico and Venezuela, hunting tuna in the eastern Pacific, invested in purse seine nets rather than in more expensive nets with wider mesh but stronger fibre. Their decision led to dolphins being trapped too, and tuna exports from these countries to the US were then restricted under US legislation designed to protect mammals. Thus arose one of the most important trade disputes in modern times: one which has raised the fundamental issue of how far it is possible to reconcile freedom of trade with higher environmental standards in a world where countries have different preferences or capacities for environmental management.

A second strand in the argument originates in parts of the business community which are deeply concerned about the way in which national environmental policymakers are acting in isolation from their competitors overseas. This leads, they argue, to 'unfair' competition or 'environmental dumping'.[1] The significance of added costs will depend on their magnitude and also on market structure which will determine how easy it is to pass on costs without losing market share. For most economies, taken as a whole, problems of 'competitiveness' may not be a serious issue since adjustment will – more or less smoothly – shift resources into other sectors which are more competitive. An OECD study suggested that pollution control may have reduced the exports of the US, France and the Netherlands by only between 0.5 and 1 per cent (though controls have tightened since the mid 1980s when the study was made).[2]

But for particular industries or firms there may well be serious concern. The European chemical industry claims that environmental expenditure now accounts for almost 20 per cent of its production costs.[3] For many energy intensive industries in Europe, the European Commission's energy tax proposals crystallized these concerns. The chairman of Bayer recently warned that his company would be forced to switch a large part of its operations overseas:[4]

> We cannot compete successfully against intense global competition ... while at the same time shouldering the burden of constantly rising levies ... (and regulations) ... If the present proposals to tax carbon dioxide emissions

---

[1] 'Environmental dumping' allegedly occurs when goods with relatively low process standards are exported at below the full cost including environmental cost. There is thus a hidden subsidy.

[2] OECD, *Environmental Policies and Industrial Competitiveness*, OECD, Paris, 1993.

[3] Author's communication with Conseil européen des fédérations de l'industrie chimique (CEFIC).

[4] 'Bayer chief warns of tax threat to chemical companies', *Financial Times*, 27 November 1991.

and solid waste were implemented, it would no longer be economic to manufacture inorganic pigments and organic intermediates in Germany'.

Some businesses and environmental lobbies have taken their concerns one step further. A US food packager, Teepak, backed a Congressional Pollution Deterrence Act allowing imposition of countervailing duties on imports from countries with lower environmental standards. Attempts will be made to use the NAFTA side letters and any new environmental provisions within the GATT to stop such environmental 'dumping'.

There is an emerging coalition of beleaguered businesses, protectionist legislators and environmentalists. The coalition is broadened in practice to include rich country labour unions whose concern over low wage competition – 'social dumping' – parallels that of environmental NGOs. The potential of such group-ings for causing lasting damage to the international trading system is serious.

2. Does trade damage the environment?

Some environmentalists think so. In particular, GATT is seen as a conspiracy against the environment. As a leading US activist put it:[5]

The World Trade Organization [which will succeed GATT in 1995] would be run by unelected bureaucrats ... This new, permanent bureaucracy would have great power ... yet would be unaccountable, undemocratic and would operate in secret. The laws at risk include vital consumer and environ-mental provisions ... Under the proposed agreement these types of meas-ures are viewed as 'non-tariff trade barriers'.

At root, the environmentalist worry over trade centres on a scale effect: if trade and investment liberalization causes expanding economic activity, there will, other things being equal, be more pollution. This could be seen as an argument against growth *per se*, originating in concerns that environmental thresholds are already being crossed. This argument is extended to develop-ing countries. It is said that these countries are forced by the international economic system, and by multinationals in particular, to 'over-exploit' their natural resources, leading to environmental degradation and to environmental

---

[5] Ralph Nader, 'Trade Action Kit', Public Citizen, Washington DC, 1991.

costs from manufacturing in a largely unregulated situation. The implication often drawn is that some withdrawal from the world economy is called for.[6] While these views do not (yet) represent the mainstream of environmentally aware public opinion, they almost certainly motivate many activists.

The mainstream of environmental thinking, however, is characterised by very different ideas on 'sustainable development', as originally set out in the 1987 World Commission report, *Our Common Future*.[7] The argument, here is that the expansion of trade through liberalization reinforces sustainable development. The Rio Summit echoed this approach. Chapter 21 of Agenda 21 opens by stating that 'the removal of existing distortions in international trade is essential'. It goes on to note that 'an open, multilateral trading system makes possible a more efficient allocation and use of resources and thereby contributes to an increase in production and income and to lessening demands on the environment'. It advocates the need 'to halt and reverse protectionism in order to bring about further liberalization and expansion of world trade, to the benefit of all countries, in particular the developing countries'.

This sense of a synergy rather than an incompatibility between trade and environmental objectives has several specific supporting arguments:

- Trade is a motor of economic growth, and particularly so in the case of outward looking developing countries. Rising living standards should reduce the poverty which lies behind much – mainly rural – environmental degradation.[8] In particular, industrial pollution intensity per capita appears to fall as income rises.

- Liberal policies in general, including open trade regimes, are associated with a lower degree of pollution than exists in comparable countries with less liberal policies.[9] One reason is that in a deregulated environment

---

[6] T. Lang and C. Hines, *The New Protectionism: Protecting the Future against Free Trade*, Earthscan, London, 1993.

[7] United Nations World Commission on Environment and Development, *Our Common Future*, United Nations, New York, 1987 (also known as 'The Brundtland Report').

[8] The World Bank, *World Bank Development Report 1992*, Oxford University Press, New York, 1992, Figure 2.

[9] N. Birdsall and D. Wheeler, 'Trade Policy and Industrial Pollution in Latin America: Where are the Pollution Havens?' in Patrick Low (ed.), *International Trade and the Environment*, World Bank, Washington DC, 1992, pp. 159-167.

resources such as water or timber are more realistically priced; there is less subsidization of energy, fertilizer and pesticides; and there is freer access to environmentally clean technology. What has happened in Eastern Europe underlines the grim implications for the environment of industrial development when isolated from the mainstream of trade among market economies.

• Where trade liberalization reduces the barriers to labour intensive manufactures and services, this enables developing countries to diversify more quickly from commodity exports, where there is some indication that unsustainable resource use takes place (examples being logging in Southeast Asia or cotton in the Sahel). The diversification of Indonesia from over reliance (in environmental terms) on timber and oil to textiles and electronics is noteworthy in this respect.[10] The changing composition of trade emphasizes the dominant role of differences between countries in factor abundance and technology. Differences in pollution abatement costs are subsidiary and will tend to be swamped by the beneficial results (for environmental emissions) of changing product composition towards labour intensive goods and away from resource intensive and possibly environmentally damaging goods.

• The Uruguay Round has had as one of its primary achievements the beginnings of a concerted attack on the large-scale subsidization and protection of commercial agriculture in rich countries. Broadly speaking, agricultural liberalization makes environmental and economic sense.[11]

On a more concrete level, there are many specific instances where increased trade helps to reconcile economic and environmental objectives. (See Box.)

It is not, however, helpful to see the debate in over-simplified terms as trade versus the environment. There may be cases where trade damages the envi-

---

[10] R. Repetto, *Trade and Environment Policies: Achieving Complementarities and Avoiding Conflicts. Issues and Ideas*, World Resources Institute, Washington DC, 1993.

[11] K. Anderson and R. Blackhurst (eds), *The Greening of World Trade Issues*, Harvester Wheatsheaf, London, 1992; E. Barbier, 'Managing Trade and the Environment', *World Economy*, Vol. 14, 1990; J. McNeil, P. Winsemius and T. Yakushiji, *Beyond Interdependence: the Meshing of the World's Economy and Earth's Ecology*, Open University Press, London, 1991.

How environmental controls generate trade: an example

California provides an illustration of the linkages between environmentalism and trade. Refining is close to capacity limits. Oil companies are being strongly discouraged from making new investments by the high capital costs associated with:

- **Reformulated gasoline**
- **Stricter regulator emission controls**
- **Stricter planning regulation for refining in the state.**

The market response is for distributors to import products from overseas refineries which have spare capacity and the technical ability to meet US product standards. These happen to be in Japan.

From an economic and environmental standpoint, the outcome is beneficial. Californians achieve environmental improvements (less refining on the coast) and continue to receive oil products (at slightly higher cost). Japan gains from its exports. Scarce capital is freed to be put to better use by oil companies. Environmentalism and free trade go hand in hand.

---

ronment and where policy intervention may be called for to improve welfare. Efficient trade assumes a global market where prices reflect costs. Environmental costs are often not reflected in market prices. This could lead to trade based on production which is environmentally damaging. Liberalization of trade can then make a bad situation worse by magnifying the distortions – the scale effect. A priori there is no way of knowing whether the negative scale effect will be outweighed by the composition effect, or by other factors including technological change. There are some specific concerns.

- If trade liberalization leads to harmonization of product standards, in the interests of easier commerce, this could expose higher standards – on for instance, food additives or pesticides – to be challenged as trade barriers. Particular acrimony surrounds the Codex rules, which, it is argued, have been 'levelled down' below what they should be to reflect the full cost of damage caused.[12]

- One effect of trade liberalization may be to encourage the production of goods in areas where environmental externalities are not taken into account (i.e. an environmental subsidy is paid) and thus undermine the efforts of other countries to price externalities fully – through the 'polluter pays' principle – or to allow for resource depletion.

- Attempts to achieve solutions to genuinely global environmental problems through global agreements could be undermined by 'free riders' if the trade regime does not allow discriminatory measures to be employed as a last resort.

The rather abstract arguments may actually be reduced to some more specific complaints about the workings of the GATT. In principle, GATT's Article 20 does provide for a balancing of trade and environmental interests:

... subject to the requirement that such measures are not applied in a manner which would constitute a means of arbitrary or unjustifiable discrimination between countries where the same conditions prevail, or a disguised restriction on international trade, nothing in the Agreement shall be construed to prevent the adoption or enforcement of any constricting party of measures: ... (b) necessary to protect human, animal or plant life or health ... (g) relating to the conservation of exhaustible natural resources if such measures are made effective in conjunction with restrictions on domestic production or consumption.

Disagreements have arisen not because of the GATT Article itself (though some environmentalists do not like the emphasis on non-discrimination) but because of its interpretation in subsequent rulings:

- The insistence that 'necessary' should also be interpreted as 'the least GATT-inconsistent' of all available GATT measures (which could mean

---

[12] 'Codex Alimentarius Commission' was set up in 1962 jointly by the United Nations Food and Agriculture Organization (FAO) and the World Health Organization (WHO) to ensure fair practices in international trade and to protect the health of consumers by established guidelines for the control of food quality and safety.

that command and control measures would not be acceptable, if a tariff or tax were feasible).

- The insistence (an outcome of the tuna–dolphin dispute) that the concern over exhaustible natural resources in subparagraph (g) cannot apply extra-territorially. While most environmentalists would argue that purely domestic environmental decisions should remain subject only to domestic jurisdiction, there is a large grey area where possible global and regional spillovers are involved (concerns, for instance, that deforestation in country X might affect the climate in country Y). And some environmentalists do not accept national sovereignty as a defence.

- The requirement that trade measures should only be used 'when all other options have been exhausted, through the negotiation of international cooperative agreements'.

- The fact that rule-making by panels takes place without the public participation and the judicial review required domestically.

In these rather subtle ways, it is argued, environmental concerns are not fully taken into account.

3. Does environmental regulation damage trade?

Environmentalists fret about the apparent indifference of international trading rules to environmental concerns. Free traders worry that environmentalism, operating either through trade policy instruments (tariffs or quotas) or through standard setting, will open the door to protectionism in a new guise. Table 1 shows how environmental measures can have far-reaching trade and investment implications. Though these measures are not, for the most part, protective of domestic producers, they could have this effect in practice, and several have already led to trade friction. The Box describes the case of Danish soft drink producers who managed to exclude competing imports as a result of a government requirement that imports be in refillable bottles, leading to an (unsuccessful) European Commission action to uphold freedom of trade.

In the past, before the EU's priority shifted towards restricting trade in tropical timber, it attempted to use GATT to try to overturn a ban by Indonesia

Table 1 Examples of environmental measures which affect trade and investments

Measures	Trade and investment impact
Sanctions against breach of international agreement (e.g. CFCs under the Montreal Protocol)	Restrictions on import/export of environmentally sensitive items and products containing them Incentives for trade in benign substitutes Transfer of technology through foreign investment.
Enforcement of high environmental product standards in some countries (e.g. Danish disposable bottles; bans on dangerous insecticides)	Imports discouraged from countries with lower standards Incentive to export sub-standard products to countries not requiring observance Indigenous firms develop comparative advantage in high standard products.
Enforcement of high environmental standards in some countries (e.g. chemicals, petrochemicals, refining standards in US, EU)	Incentive to import from lower cost sources, unless imports restricted Incentive to relocate to low standard countries Indigenous firms develop comparative advantage in high standard process equipment.
Conservation of depleting resource or threatened species via export restrictions or tax (e.g. raw logs; CITES)	Incentive to value added export industries Smuggling.
Restrictions on trade in toxic wastes (e.g. Basle Convention)	Legalizes trade only where adequate facilities Incentive to destroy waste at source Some evasion (e.g. deep sea dumping).

How environmental controls can stifle trade

In 1986, Denmark was brought before the European Court over a law requiring soft drinks to be sold only in returnable bottles, with a compulsory deposit. The European Commission argued that environmental protection was disproportionate and also protectionist. Denmark won. The judgement has encouraged other EC countries, notable Germany, to press ahead with comparable, or more onerous, measures involving obligatory recycling, or taxes on non-recycling, for bottles and other packaging. Why might this affect trade?

- Higher transport costs from more bulky or heavier packaging (bottles rather than cans).

- The implication is that to carry out business in a country, an exporter must set up a local (loss making) operation for deposit collection and to oversee recycling or other environmentally efficient disposal.

- Collusion between government standard setters and local firms can give the latter a competitive advantage by 'tilting the playing field' to suit local suppliers.

---

on the export of raw logs (which it believed was a disguised form of protection of Indonesia's wood processing industry). More recently, GATT ruled that the use of the US law protecting dolphins to restrict Mexican tuna imports was illegal, because it may have been a device for protecting US fishing fleets. Moreover, GATT ruled that 'a contracting party may not restrict imports of a product merely because it originates in a country with environmental policies different from its own'. In practice, therefore, many environmentally driven trade restrictions breach GATT rules.

At the heart of the GATT concern is the fear that – intentionally or not – environmental trade measures discriminate between national and foreign producers. This is not to say that agreements should cease to be concerned with environmental conservation. GATT simply wishes to emphasize that the 'first best' solution is rarely a trade restricting measure, but usually amounts to a non-discriminatory tax on all consumption or production, and not simply on imports or exports. Thus Indonesia, and other timber producers, should tax or restrict log production, for use in domestic as well as foreign timber mills.

However, even if GATT rules are faithfully observed, there are still areas where 'good environmentalism' and freedom of trade may conflict; and will do so increasingly in future. Particular conflict may arise in the following areas:

- Enforcement of international environmental agreements governing global concerns;

- Enforcement of product standards in one country which are higher than in others;

- Enforcement of production standards in one country which are higher than in others, as a result, for example, of a 'life cycle' approach to product standards, taking account not just of the final but the component inputs and processes used in its production.

These are considered in turn below.

4. How can international environmental agreements be enforced?

At first sight, one of the clearest cases for allowing trade to be managed in the interests of the environment is where there is an international conservation agreement concerning a global environmental externality, where most countries subscribe and where the agreement is threatened by 'free riders'. In several cases, trade measures have been used to underpin the agreement. For example, Japan and Norway have prohibited whale imports from non-signatories to the Whaling Convention. A similar prohibition is threatened by the Montreal Protocol on chlorofluorocarbons (CFCs) which allows action against the trade with non-signatories in banned substances and will ban products (e.g. refrigerators) containing controlled substances from non-signatories.

Most free traders now accept that a change does have to be made to GATT rules to make explicit the right of countries to restrict trade in goods that create global or cross-border environmental damage or that use processes which do this.[13] This would secure the basis of agreements like the Montreal Protocol and the Basle Convention. These grounds, like those already ceded under GATT, then have to be qualified by clear safeguards along the following lines:

- *Non-discrimination* National enforcement must be non-discriminatory between domestic and foreign firms and must respect GATT principles of national treatment and transparency. There should be no protectionist intent designed to favour domestic producers.

- *Sound science* Agreements should be based on sound science. In key areas like global warming there are areas of continuing scientific doubt.

- *Instruments* Policy, where possible, should rely on economic instruments which are more transparent and economically efficient than regulations.

- *Consensus* Sanctions are a last, not a first, resort. 'Free riders' may be countries which initially do not accept the global agreement. Attempts to impose trade sanctions could then lead to conflict and/or to continued evasion. 'Free rider' problems would be even more acute if concerted action were ever taken on global warming. OPEC has already voiced objections to the EU's proposed (non-discriminatory and GATT-friendly) energy tax. To be accepted, a trade measure would have to be part of an equitable and efficient approach by contracting parties to a common environmental problem.

- *Proportionality* The measures should be proportionate to the problem addressed. One of the basic problems with environmentalism, however, is that many judgements cannot be quantified in cost-benefit terms. How should the preservation of whales or elephants be set alongside wider trade interests? Cost-benefit rules are difficult enough to apply domestically, let alone internationally, where there are suspicions of bad faith and protectionist intent.

While trade measures may be a necessary underpinning for global agreements, they have to be very carefully designed to avoid being perverse in their effects. A classic case arises with ivory: an exotic example but one with

---

[13] J. Bhagwati, 'Trade and the Environment', *American Enterprise*, May/June 1993.

broader lessons. Official export restrictions (as by Kenya) merely had the effect of restricting supply and increasing the price, so increasing the incentive to poach. Recently attention has switched under the Convention on International Trade in Endangered Species (CITES) agreement to banning all trade. Because of the reduction in import demand in primary markets like Hong Kong, prices have fallen, reducing poaching. But there are two other consequences. First, illegal trade (to less controlled markets) now dominates the (smaller) market. Second, incomes have fallen in those producer countries (like South Africa, Botswana and Zimbabwe) which were managing elephants sustainably, so reducing their incentive to continue to do so. The final outcome of the CITES treaty is not yet clear, but there are deep divisions among conservationists as to whether a trade ban is the best way to proceed.

As pressures mount to impose global norms, there are two major lessons from experience to keep in mind:

- Trade will not take place without demand. Outlawing trade is invariably less efficient than acting directly on demand through non-discriminatory taxes, education or information, whether the demand is for drugs, ivory, tropical timber, whale meat, dangerous chemicals or non-degradable packaging.

- As long as demand and supply exist it is better to legalize (and manage) the international market through agreed rules – as in a commodity agreement – than to outlaw trade. Internationally traded pollution (or depletion) permits would be one way of achieving this objective.

5. Are higher environmental product standards a barrier to trade?

Trade is often restricted in order to uphold environmental and public health and safety standards. Environmental grounds are the most common reason cited for trade restrictions at the border in order to enforce these standards (accounting for 55 per cent of cases between 1980 and 1990 notified to GATT). The most common are emissions controls on cars, ozone depleting substances and toxic chemicals. India's exports of leather to Germany, for example, were recently stopped by a ban on polychlorinated bi-phenol (PCB) preservatives until a substitute was found. More generally, German ordinances on cutting packaging waste and the Danish reusable bottling requirements discussed earlier have forced exporters to reorganize their marketing activities.

Trade has also been affected, indirectly, by shifting consumer tastes, especially when reinforced by 'eco-labelling'. The current approach to labelling in the EU (Regulation 880/92) deals not just with the product itself but with its whole lifecycle, in Europe and overseas. Pulp and paper exporters in Brazil, for example, have expressed concern that they face hidden barriers to trade from eco-labelling. Tropical timber imports have been similarly affected; NGOs in the UK, for example, have tried to set standards for tropical timber used by councils in local authority construction projects.

At first sight, there should be no problem with countries raising their own environmental product standards and requiring importers to meet them, on a non-discriminatory basis. GATT accepts the right of countries to improve environmental product standards. There is a Standards Code which reflects this principle and which tries to ensure that standards are clear ('transparent' in GATT jargon) and are not being used as a pretext for import restrictions (in the way that Japanese anti-pollution standards have allegedly been used to keep out foreign cars and the United States Corporate Average Fuel Economy (CAFE) standards may be becoming a barrier to imports from the EU and elsewhere).

There are, however, some difficulties in this area: one specific and one more general. The specific problem is whether exporters should be required to observe the same norms abroad as they do at home. This issue arises in relation to the export of dangerous chemicals (e.g. insecticides), pharmaceuticals, baby foods, and hazardous waste. To some extent, problems are the result of genuinely different standards in overseas markets. This raises the question of whether exporter governments and companies have a right to impose, and charge for, quality standards which the customer may not seek: and further, whether they should have a redress against foreign competitors who ignore those standards. Problems can also arise because of ignorance, raising the question of how much responsibility the supplier has to inform the consumer of environmental and safety considerations. Efficient international trade requires the maximum disclosure and dissemination of information. This policy, rather than trade controls, is the best long-term safeguard of standards. Indeed, this is the basis on which governments and chemical companies operate the International Register of Potentially Toxic Chemicals (IRPTC).

The more general point is that different nationally imposed standards may act as a barrier to a deeper integration of markets. For this reason, the EU gives high priority to harmonizing environmental and other standards. The GATT's Codex rules are also designed to achieve common standards. Harmonization, however, opens up potential conflict where some countries feel their standards are being diluted and others feel that the standards are so onerous as to represent a barrier to trade. Some degree of conflict over standards is inevitable, which is why it is crucial to have mechanisms for dispute settlement which can balance the interests concerned, ensure transparency and prevent standards being used for protective reasons.

6. Is environmental 'dumping' acceptable?

A related issue arises where different environmental standards for production processes result in firms in the high standard country being made uncompetitive and losing out to 'unfair' 'cheap' imports which do not incorporate the environmental costs of production, i.e. which enjoy an environmental 'subsidy'. The issue is given added significance by the demands increasingly being made to apply a 'lifecycle' approach to environmental product standards.

In practice, there has been strong resistance in the GATT to allowing different environmental process standards to be used as a basis for such trade action. The ruling on US–Mexico tuna trade was a major example. Several arguments have been advanced:

- Strict environmental standards impose a higher real cost on poor countries where capital is scarcer and where environmental costs may genuinely be valued much less, relative to the benefit of higher environmental standards. It is very unlikely, for example, that any rational calculation of cost and benefit would lead many poor countries to control sulphur emissions as severely as Europe, Japan or the US. More controversially, workers in highly polluting industries, in poor countries, may choose to work and suffer injury in these conditions because alternative employment options – if any exist – are worse.

- In some countries there may be greater capacity to absorb environmental damage. This is clearly a dangerous argument, open to abuse, but since much environmental concern is about crossing thresholds, it is legitimate to argue that not every case should be judged as if the country were crossing the threshold (it may, for example, have relatively lax fishing controls because its coastal waters are generously stocked; and smoke stack emission standards may sensibly be lower in areas with low population density).

- The argument about 'competitiveness' is a static one. Just as manufacturers in high wage countries are not permanently handicapped relative to low wage producers, because of the spur to productivity, so high environmental standards encourage producers to diversify and to develop a competitive advantage in low polluting technology. The success of Japanese and German environmental equipment exporters is testimony to this process.

Experience has shown that, in practice, 'environmental dumping' is not a major influence on trade. Many other factors are involved.

This particular problem would disappear if all countries observed the 'polluter pays' principle (PPP) since environmental costs would then be reflected in prices. In principle, exporting countries could then reasonably be expected to justify environmental 'subsidies' (the non-observance of PPP) and, as a final resort, countervailing tariffs could be available as a defence where domestic producers are damaged. However, in practice, the PPP is rarely observed; the practical problems of calculating environmental subsidies (defined differently in different countries) would be immense; and the history of anti dumping policy is one of perpetual abuse on protectionist grounds. Differences of approach suggest that environmental process standards are likely to be a major battleground in trade policy, but that there are good reasons for resisting greater freedom to use trade instruments.

7. Foreign investment flows and the environment

We have concentrated on the international economic ramifications of environmentalism from the standpoint of trade. However, trade and investment flows are interconnected, a fact belatedly recognized in the GATT (which is drawing up an agreement to restrict trade distorting foreign investment measures).

Environmental regulation can affect investment flows by driving up manufacturing costs in environmentally sensitive countries. Regulations and taxes provide an incentive to move 'offshore', to 'pollution havens', both to gain a competitive advantage over less mobile firms and to defend a competitive position against other companies in the 'havens'. The process is analogous to (indeed part of) that of environmental 'subsidization' referred to above – in this case 'subsidization' to attract foreign investment rather than (or as well as) exports.

A good deal of research has now been carried out as to how important the phenomenon is in practice. There has undoubtedly been some migration of chemical and non-ferrous metal processing overseas because of environmentally driven high cost standards related to emission controls, waste disposal or planning regulations. A study by Leonard showed that in two industries (chemicals and metal processing) there had been some migration to four countries (Ireland, Spain, Mexico and Brazil) which offered low environmental standards – in some cases, like Ireland, explicitly so.[14] Other studies point to the same conclusion. There was some relocation of copper smelters and refineries, asbestos, ferro alloy and vinyl chloride plants a decade or more ago and perhaps a disproportionate number of petrochemicals or chemicals plants relocated in southern Europe, from Germany and the Netherlands.[15]

But, in general, the relocation effect is weak. Moreover, only in very few cases are environmental costs a decisive factor in location decisions. Pollution 'havens' have themselves started to raise standards, forcing foreign investors to move or close down (cases from Taiwan and Ireland are cited). Where market growth in the industry is strong, most companies will prefer not to migrate but to undertake expensive pollution controls or develop substitutes knowing that costs can be passed on in these markets (though, arguably, this comforting possibility is no longer present in the depressed and competitive chemicals and metals markets of the OECD).

Indeed, there is some indication that foreign investment may flow in the opposite direction from that suggested by the existence of pollution 'havens'. Where large companies have acquired a corporate commitment to high environmental standards there is reluctance to dilute (or be seen to dilute) these in countries which neither expect nor enforce high standards. Thus there

[14] Jeffrey Leonard, *Pollution and the Struggle for World Product*, Cambridge University Press, Cambridge, 1988.
[15] M. Rauscher, 'Foreign Trade and the Environment' in H. Siebert, *Environmental Scarcity: The International Dimension*, Mohr, Tübingen, 1991.

may be a choice between making a costly, quality raising, investment in an unattractive business environment or withdrawing from the country altogether. This is a particularly live issue in Africa.

8. The EU energy tax proposal and 'competitiveness'

The problems presented for international trade by environmental measures in relation to timber, CFCs and whales might pale into insignificance beside those which will arise should governments elect to take significant preventative action to reduce greenhouse gas emissions.

At present, policy consensus centres only on 'no regrets' measures which, by the most common definition, have few costs. But these are unlikely to suffice for the more ambitious targets which have been set. The current topical policy issue is the possibility of carbon (or general energy) taxes. There is a large range of estimates of the level of tax which would need to be levied to achieve current (Toronto) target levels, depending on the supply and demand responses employed in the model, but a substantial tax on primary energy – totalling 50 to 100 per cent, if not more – would probably be required.[16]

The recent EU Commission proposal to introduce a combined energy/carbon tax is consistent with those estimates and brings into sharp focus several of the issues raised above. In many respects, the EU tax proposal is a model of how to avoid trade disputes consequent upon environmental measures: it is non-discriminatory as between imports and domestic producers; it is transparent; it is market-based rather than regulatory; it has no protectionist intent (indeed, most of those who feel aggrieved are domestic energy producers and consumers); and it involves voluntary action to tackle a global problem rather than imposed discipline on 'free riders' (the rest of the world, in this case).

Despite these credentials, the tax (or taxes) has met with substantial opposition. One strong objection to the tax is that it puts European producers at a competitive disadvantage, because the EU has high standards even though global warming is an internationally shared problem. This competitive disadvantage is not a significant problem for European primary energy producers since competing imports of coal, oil or gas would be taxed at the

---

[16] J. Whalley, 'The Interface between Environment and Trade Policies', *Economic Journal*, No. 101, March 1990.

point of entry and their exports would be rebated. The concern is for energy intensive industries.

The main impact is felt by a small number of industries: ferrous and non-ferrous metals, chemicals (though the impact is reduced if chemical feedstocks are exempted, the feedstocks being transformed into carbon based products, not carbon dioxide), and non-metal minerals (mainly cement and glass, where transport costs limit extensive trade). The major unresolved problem relates to metals and bulk chemicals, especially as they are homogenous commodities for which price is the key factor in competitiveness.

How could the problem of competitiveness be dealt with? There are various options:

* One approach is to recognize that if the EU does want to act as an environmental leader in this case, costs have to be paid by someone: producers or consumers. It is, according to this argument, appropriate that some of the cost be borne, in the form of lost markets, by energy intensive industries to the extent that they cannot adjust to less carbon- or energy-intensive methods.

* Another approach is to exempt the affected industries altogether. This is the route currently preferred; steel and non-ferrous metals, chemicals, cement, glass, and pulp and paper are promised exemption if the tax is introduced. This has the unsatisfactory effect of giving no energy/carbon reducing incentive to such industries and, perversely, of benefiting these energy-intensive industries while taxing others.

* A third approach, which is analytically more satisfactory but perhaps too cumbersome in practice, is to treat the tax like another major EU indirect tax, VAT, which is levied on imports but from which exports are exempt. Thus, energy-intensive industries could claim energy tax exemption on their exports as they do with VAT on their other inputs. There is a special problem with competing imports. As a solution, imports could be taxed on the energy content of their production as, for example, the EU currently does with the sugar content of tinned fruit.

The problem of competitiveness disappears if the tax is applied multilaterally, which is a prior requirement for introduction of the EU tax (at least in relation to the OECD). However, problems of a different kind might then arise. Compliance costs vary from one country to another (US output is 40 per cent more carbon intensive than that of the OECD as a whole; the German is higher than the French). Either tax rates would have to vary to achieve the same result, or the rates would be the same but have different results: both systems providing fertile grounds for complaints of 'unfairness'. Moreover, non-OECD countries would not be covered and this raises the question of the length of time big carbon emitters such as China or India would be allowed to remain as 'free riders' if they refused to join a collaborative approach. The Montreal Protocol gives some hope that, with the help of compensation payments, these problems can be resolved globally, but the CFC problem is trivial in size by comparison.

The problems associated with a carbon tax, notably the trade–competitiveness concerns, have prevented it from being agreed. But the problem will not go away. Taxes are not the only market instrument under consideration. Tradeable permits have the additional advantages of being precise in their impact and providing reluctant participants with compensation.[17] (The same mechanism has been used to coax textile exporters to agree to trade restraints.) But governments are a long way from agreeing the basis of an allocation.

The important issue to establish at this stage is that, if the principle is accepted for acting on global environmental problems, this should be done in ways which do not create trade barriers that discriminate against producers – carbon taxes meet that test – and make maximum use of non-trade measures (such as compensation payments or traded permits). If trade restrictions are used as a last resort they must be surrounded by safeguards of the kind sketched out in Section 4 above.

9. Conclusions

In the 1990s environmental issues will increasingly spill over into trade policy (and vice versa). Trade disputes based on differing environmental standards or enforcement of agreed standards can be expected to multiply. Either a

---

[17] Michael Grubb, *The Greenhouse Effect: Negotiating Targets*, RIIA, London, 1989 (second edition 1992).

disciplined application of trade measures for environmental ends will have to be developed in a GATT framework, or anarchic conditions will prevail, influenced by unpredictable environmental pressure groups and coalitions with protectionist interests.

Environment and trade is an area where there is enormous scope for misunderstandings, suspicions of bad faith and dubious motives. But there are no simple formulaic solutions. There are undoubtedly cases where unrestricted trade would increase environmental costs; but whether trade restrictions would be a satisfactory 'second best' solution depends on a complex balance of costs and benefits and consideration of alternatives. The 'polluter pays' principle is a good theoretical basis for reflecting environmental externalities in trade, but would be difficult to measure and monitor to the satisfaction of disputing parties.

Looking ahead, an underlying question is whether there should be a general bias towards encouraging *harmonization* of national environmental standards, or instead a looser, *mutual recognition* approach to rules and standards. This fundamental divergence of approach has dogged attempts at regional integration in the EU (more recently in the North American Free Trade Agreement) and will dominate any future GATT negotiations. The GATT tradition is one of mutual recognition (and this has also emerged as the guiding principle of the EU). But there are pressures for harmonization both on grounds of simplicity and business efficiency and also because of concerns over 'fairness' and 'level playing fields'. Environmentalists tend to adopt a contradictory position, arguing for harmonization of process standards, or at least agreed minima – to avoid 'environmental dumping' – but against harmonization of product standards, for fear of 'dragging down' the highest standards. Anything other than a voluntary approach to standards harmonization at a global level is likely to lead to confusion and conflict.

If the environment is to play a more weighty role in international trade negotiations there will be pressure to create new global initiatives to monitor and develop policy in relation to trade and the environment. The key issue here is what happens to the proposed World Trade Organization (WTO). It will, in principle, make the GATT much stronger. It will also change the dispute settlement procedure to give the resolutions of panels (like that of the tuna–dolphin dispute) more force. In addition, the burden of proof will be on the defendant (who imposes the trade barriers) and unilateral trade restrictions will be policed. By and large, weaker (developing country) parties welcome the WTO since it heralds the approach of a rule-based system. Environmentalists worry about it, especially in view of the history of GATT panel rulings, but are working hard to ensure that their role is strengthened. Either way, the tension between trade and environmental concerns is real and will grow.

In conclusion, it may be useful to review the various ways in which this tension can be lessened. There are environmentalists who are worried at the way more militant colleagues could damage the trading system. And there are economists who see that cases exist where total freedom of trade might have negative environmental side effects in some instances. Common ground would include:

- *Transparency* There should be a mechanism to look at the transparency of trade-related environment measures along the lines of the GATT Trade Policy Review Mechanism, which already exists. At the same time there is no excuse for GATT panels to meet behind closed doors; discussion should be open.

- *Global environmental hazards* Some provisions to legitimize trade action, as a last resort, against free riders in international environmental agreements will be necessary to underpin agreements like the Montreal Protocol.

- *Safeguards on safeguards* Where trade restricting measures are allowed, to reinforce international environmental agreements or to enable countries to uphold their own domestic product standards, there must be clearly defined safeguards including: an absence of protectionist intent; non-discrimination between foreign suppliers and also between domestic and foreign suppliers; sound science; and a test of reasonableness. These crucial details will, however, be enormously difficult to negotiate in practice.

While these issues are all potentially negotiable, it is also possible that the trade–environment issue could become seriously divisive, particularly if attempts are made at forced harmonization of process standards to limit what is described as 'environmental dumping' from low cost suppliers.

General references

C. Arden-Clarke, *South–North Terms of Trade: Environment Protection and Sustainable Development*, WWF Discussion Paper, Gland, Switzerland, 1992.

W. Baumol, *Environmental Protection, International Spillovers and Trade*, Almquist and Wicksell, Stockholm, 1971.

J.N. Bhagwati, 'Trade and the Environment: The False Conflict?' in D. Zaelke, P. Orbuch and R.F. Housman (eds), *Trade and the Environment: Law, Economics and Policy*, Island Press, Washington DC, 1993, pp. 159-190.

F. Cairncross, *Costing the Earth*, Business Books, London, 1991.

J. Cameron and H. Ward, *The Multilateral Trade Organization: A Legal and Environmental Assessment*, WWF, Gland, Switzerland, 1992.

J.M. Dean, *Trade and the Environment: A Survey of the Literature*, World Bank, Washington DC, 1992.

Environment Law, Issue on trade and environment, Vol. 23, No. 2, 1993.

H. French, *Costly Trade Offs: Reconciling Trade and the Environment*, Worldwatch Institute, Washington DC, Worldwatch Paper 113, 1993.

General Agreement on Tariffs and Trade, *Trade and the Environment*, GATT, Briefing No. 1529, 1992.

G.M. Grossman and A. Kreuger, *Environmental Impacts of a North American Free Trade Agreement*, Centre for Economic Policy Research Discussion Paper No. 644, London, 1992.

G. Hansson, *Harmonization and International Trade*, Routledge, London, 1990.

C. Hines, *Green Protectionism*, Earth Resources Research, London, 1990.

G. Hufbauer and J. Schott, *NAFTA: An Assessment*, Institute for International Economics, Washington DC, 1993.

Journal of Environment and Development, Issue on trade and environment, Vol. 1, No. 2, 1993.

P. Low (ed.), *International Trade and the Environment*, World Bank, Washington DC, 1992.

OECD, *Environmental Policies and Trade Competitiveness*, OECD, Paris.

C. Pearson, 'Industrial Relocation and "Pollution Havens"', *Economic Impact*, No. 65 USIA, 1988.

R. Repetto, *Review of European Community and International Environment Law*, (Trade and Environment Issue), Vol. 1, No. 1, 1992.

D. Robertson, 'Trade and Environment: Harmonization and Technical Standards' in P. Low (ed.), *International Trade and the Environment*, World Bank, Washington DC, 1992, pp. 309–321.

S.J. Rubin and T. Graham (eds), *Environment and Trade*, Allanheld Osmun, New Jersey, 1982.

D. Runnalls and A. Cosby, *Trade and Sustainable Development: A Summary*, International Institute of Sustainable Development, London, 1992.

I. Walter, 'Environmentally Induced Industrial Relocation to Developing Countries', in *The World Economy*, Symposium on Trade and Environment, Vol. 15, No. 1, January 1992.

Part 3

Energy Industry
Experience and Perspectives

Impacts on European Oil Industries

*John V. Mitchell, Chairman, Energy and Environmental Programme,
Royal Institute of International Affairs and Former Special Advisor
to the Managing Directors, British Petroleum, London*

Politics, Economics and Environment:
Experience of the US Oil and Gas Industries

*Clement B. Malin, Vice President, International Relations,
Texaco Inc, New York*

Sustainable Coal Use: the Role of Technology

*Anthony Baker, Head of Economics, British Coal Corporation, London,
and
John C. Whitehead, Head, Coal Research Establishment, Cheltenham*

The Pros and Cons of Learning by Doing:
the California Utilities' Experience in
Energy Efficiency and Renewables

*Fereidoon Sioshansi, Manager, Electric Power
Research Institute, California*

Impacts on European Oil Industries

John V. Mitchell

Abstract[1]

Environmental legislation and policy in Europe affects the quality of oil products for transportation, the quality of fuel oil demanded by oil consumers, the industry's own operations, liabilities for cleaning up sites, and the risks of oil spills. Objectives aiming to protect human health, preserve ecosystems and protect the climate sometimes overlap and sometimes conflict. In most cases, achieving successful environmental outcomes requires action not only by the oil industry, but by its customers and by related industries.

Product quality *Unleaded gasoline and catalytic converters are now established policies. The oil industry is reducing volatile organic compound (VOC) evaporation from its own operations, but is in debate with the European Commission and the automobile industry about whether emissions from vehicles being filled or in use should be further monitored by closely controlled systems or by additional changes in gasoline specifications. Further reductions in sulphur levels in diesel and marine fuel are also under discussion, and could prove very costly. Constraints on the use of sulphur by industry and electricity customers have already driven markets to the use of gas.*

Oil spills *Minimizing the risk of oil spills is a major issue. A quantum leap is needed to enforce existing standards on vessels and practice. Failures put*

---

[1] This paper is based in part on a presentation by the author to the IPEC/OPEC Workshop on The Impact of Energy Taxes and Environmental Measures on Consumption, Prices and Investment in the Petroleum Industry, Vienna, 22–23 September 1993.

129

the whole oil and shipping industries at risk from public reaction to oil spills.

Operations and site liability *For petroleum operations, there will be further reductions of sulphur emitted to the atmosphere from refinery fuel and of oil discharged into the North Sea from produced water and oil-based muds.*[2]

Site liability is the subject of a Green Paper for discussion with the Commission by the oil and insurance industries and the legal profession. There are continuing uncertainties on liability which inhibit investment and divestment.

Climate protection: carbon dioxide and taxes *Carbon taxes are not doing well politically. Meanwhile, the industry faces a 'mixed bag' of gasoline tax increases and threats of regulation. European governments have taken advantage of crude price reductions since 1985 to maintain or increase gasoline taxes in real terms.*

Industry attitude *Industry is responding to these developments by a mixture of reaction, argument and initiatives. Not much is being done to put the main case for oil as a fuel. Its economic value is measured by the high taxes consumers are prepared to pay. Oil's environmental impact, including its carbon dioxide emissions, is superior to alternative fuels in many applications.*

---

[2] Produced water refers to the water that is also produced when oil is produced. Invariably the water contains traces of oil and is either discharged at sea or returned to the reservoir.

Introduction

This paper has three parts: first, a review of major trends in environmental policy in Europe which affect the petroleum industry. This will be focused on three issues: product quality, transportation, and site liability.

Second, a review of trends in the climate policy debate and in consumer taxation of fuels. As I will show, the latter does not have very much to do with environmental policy.

Third, what does it all mean for the petroleum industry? In my opinion the economic weather is not very good. Some parts of the industry will get through in better shape than others, while customers and related industries will also be affected.

Trends in environmental policy

Environmental policy is a wide subject and not a simple one. It does not affect the oil industry alone, but all industrial operations – some more than others – each with a particular set of details.

I think it is helpful to analyse the process by which environmental policies are generated. The process I describe is typical, I believe, of many OECD countries, but is certainly not confined to them. It is a process which moves in one direction only. It is driven by the degradation of the natural environment under the combined influences of rapidly increasing population, and economic development which consumes and transforms natural resources – including the air, water, and perhaps the climate.

Degradation comes in many forms: some are well understood scientifically. There is not much doubt that concentrations of metals in drinking water and air, or of sulphur or organic compounds in the air we breathe are damaging to health, though there will be argument about precisely what level of concentration causes how much damage. A connection between volatile organic compounds (VOCs), low level ozone, climate, and health in certain locations is also reasonably well established. These form a *scientific input* to the environmental policy process.

There is also an input from *public opinion and perceptions*. The perception that the air in places such as Los Angeles or Mexico City was foul and unhealthy provided a powerful political motor for the adoption and enforcement of air quality standards. A particularly bad experience – a week of smog which kills a dozen people, for example – raises these perceptions to the level of outrage, which the media communicates because, as always and everywhere, outrages make news. The more the public is outraged by what it perceives, the stronger is the drive for the political process to do something: something visible. The need for visible action can be stronger than the need to know what to do or how to do it. This is not necessarily a bad thing, since some of the effects of environmental degradation are slow and cumulative: the death of the Aral Sea, for example. A little outrage may get attention for a real problem before its effects become visible, and while there is still time to take remedial action. (This is a strong argument in the climate change debate.) However, since almost any action costs someone something, the debate will not be completely objective, and action without science will be risky for everyone: those who suffer from the action, and those whose hopes of environmental benefit may be disappointed if unscientific policies are enforced. I will say more later about what this means for the role of the oil industry in the debates which affect us.

A final ingredient in the process is *idealism*. There are advocates for whom environmental policy serves a larger purpose, who will use outrages to advance their cause, and who will work unceasingly, often for no personal reward, to keep environmental issues on the political agenda. Anyone who has dealt with the so-called 'Green Parties' or Green organizations knows that ideals and methods vary between organizations. Within each movement, also, there may be conflicts between the desire to do something and the willingness to take time working out how best to do it. We should not be surprised to find contradictions within environmental movements and between different environmental programmes. Justifying any realistic idea of a better world is likely to be as complicated as justifying the status quo.

The question of *global warming* is a relative newcomer to this process, following behind the older and more obvious issues of clean air, clean water, and safe products and processes. It differs from them in other ways:

- First, CO_2, methane, and CFCs are not proven pollutants in the same sense as lead and sulphur are. The concern about CO_2 arises solely from the suspicion that its build-up in the atmosphere may lead to increases in the average temperature of the planet at some time in the next two centuries, and that this global warming could result in local and regional changes of climate which might be severe although we have no way of predicting their level of severity.

- Second, most of those who are most convinced about the existence of the problem recognize that it cannot be solved locally. One country cannot protect its own climate.

- Third, the uncertainties about the science are not going to be resolved quickly, but waiting for a better understanding may mean that the problem – if there is one – becomes significantly worse. This is a classic example of decisionmaking in an atmosphere of uncertainty, attracting the attention of a lot of clever advisers to decisionmakers.

I have described the process by which environmental policies are made. What is the result of the policymaking process?

I think of the result as a caravan of policies. The caravan moves slowly in one general direction, that of a cleaner, safer and 'sustainable' planet. There are many travellers in the caravan. Most have specific agendas, some of which are incompatible with each other and must be traded off on the way. Some are driven by broad visions of the disaster which will overwhelm us all if the caravan does not move. Some are really fellow travellers, going along for the fun, the campfire, and the adventure. One thing is certain. Moving the caravan along absorbs the resources of the communities it passes through. Most of the rewards are due in some other place, at some other time.

There are three particular components in the caravan which is moving across the petroleum world.

- The first is the cluster of policies affecting *emissions* designed to protect the quality of air and water used by the communities which use our products or host our operations. These mainly affect the products we manufacture and the way in which we control their handling and transportation on land.

- The second is the cluster of policies protecting *ecosystems*, natural habitats such as wetland or marine life. These affect the availability of acreage

Figure 1 Environmental caravan

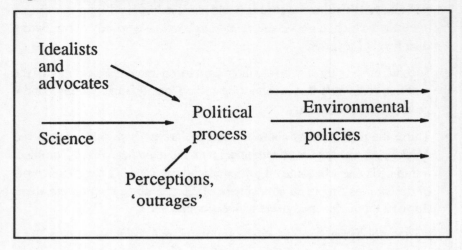

for exploration and production operations, and they affect marine transportation: oil spills are the key to the outrage that has driven these policies, but preserving biodiversity is also becoming more and more important.

- Third is the cluster of policies, still in their early days, connected with protecting the climate from the *consequences of global warming*.

The connections between these policies are complex. Many of the policies requiring cleaner products can only be met by more energy-intense refining processes. In fact, in Europe since 1980 it is probable that the whole saving of refinery fuel due to more efficient management and control has been offset by the need to use more energy-intensive processes to remove sulphur and lead from petroleum products.

Protecting ecosystems from the alleged threats of oil exploration and development has effectively closed the Atlantic and Pacific offshore areas of the US to exploration and restricted future oil production accordingly. In the long term, this means greater demand for oil from the oil-exporting countries.

The climate protection policies, if seriously applied, could have a different effect: by promoting the development of non-fossil alternative fuels and increasing investment to reduce energy consumption, oil demand might be reduced in the very long term from the new heights it would otherwise reach.

Figure 2 Mixing climate with other policies

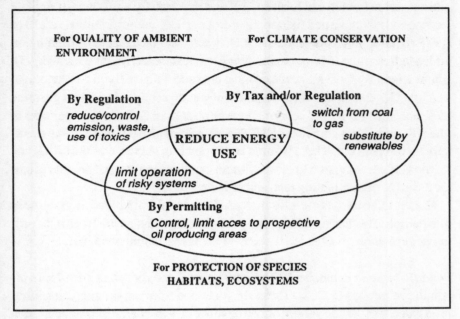

These are long term possibilities which will be difficult to achieve, and in my opinion are of less urgency for the oil industry than regulatory and permitting policies.

Let us turn to the environmental questions that I see affecting the oil industry in Europe.

Product quality
There are several policies which have the objective of protecting human health by reducing or eliminating certain pollutants created when oil products are used. Pollution by the final user is lessened by reducing or eliminating constituents from the product which the refiner delivers to the user. Examples are reduction of lead in gasoline, reduction of volatile organic compounds, reduction of sulphur, and reduction, or at least control, of particulates.

Lead There are two objectives regarding lead: to reduce the amount of lead pumped into the air by exhaust emissions and hence the amount of lead absorbed by the population; and to reduce emissions of carbon monoxide and nitrous oxide (which can form smog), and incidentally emissions of organic compounds in unburned fuel. In the European Community, Directive 210 of 1985 required that cars made since October 1990 be capable of running on unleaded gasoline (Motor octane number 85, Research octane number 95). From 1 January 1993 all new cars must be fitted with catalytic converters (to remove CO and NO_x) which require unleaded gasoline. Unleaded gasoline now accounts for 85 per cent of all gasoline sold in Germany, 47 per cent in the UK, and 48 per cent in the EU. These percentages will increase – eventually to 100 per cent. There is some corresponding decrease in the efficiency of the vehicle compared to those without catalytic converters, but that is part of the price of eliminating lead and reducing NO_x.

The oil industry has steadily increased the supply of unleaded gasoline at the pumps. This has enabled some governments, such as the British, to promote the switch to unleaded fuels by lower taxes on unleaded fuel.

Volatile organic compounds The next target is the reduction of volatile organic compounds (VOCs). These are generated from the storage, transportation and use of a wide variety of petroleum and petrochemical products (such as solvents). Petroleum accounts for about 40 per cent of Europe's VOC emissions. Of the 40, 35 per cent comes from the vehicle (nearly 25 per cent from unburned fuel in exhausts and about 10 per cent by evaporation from cars). Of the remainder, just over 5 per cent comes from storage and transportation and 2 per cent from the refuelling operation.

The EU is committed to reduce all VOC emissions to two-thirds of 1988 levels by 1999. Emissions from vehicles will be tackled by the requirement that all new cars sold in the EU must be fitted with catalytic converters, and a small carbon canister to absorb evaporative emissions while the vehicle is stationary.

A 'common position' on a further Stage 1 Directive was adopted by a qualified majority at the June 1993 Council of Environment Ministers. It is aimed at the 5 per cent of VOC emissions caused by displacement during storage, transportation and refuelling. The industry is proceeding to install double seals on floating roof tanks,[3] fit internal covers on other types of tanks, paint

---

[3] Floating roof tanks have roofs that float on top of the oil to prevent evaporation.

large tanks with reflective paint to lower temperatures, minimize evaporation during unloading by bottom discharge of tank wagons, and recover vapour from transporters while these are unloaded.

The oil industry is not committed to the so-called Stage 2 proposals which would require filling stations to fit equipment to capture and recycle vapour generated during the filling of vehicle tanks. A better alternative, in the oil industry's view, would be for the captured vapour to be absorbed, but this is opposed by the automobile industry. The oil industry argues that if the fuel system for automobiles is closely controlled to exclude any but accidental vapour emission, there is no need to constrain the content of the fuel further, for example by reducing the permitted benzene content below the current 5 per cent.

The VOC issue is also linked with the question of so-called biofuels. To an American audience, biofuel probably means firewood under stoves in houses in Maine. To a South Asian, biofuel probably means cowdung. In the European Union, biofuels mean ethanol and esterified oils produced from unsaleable surpluses of grain and rapeseed grown on French and German farms. The oil industry, supported for once by many environmental groups, has succeeded in frustrating a Commission proposal to exempt such fuels from most of the high excise taxes which apply to petroleum products. The farmers still have friends, however, especially in France, and the story is probably not over.[4]

Sulphur: the details matter The EC Council of Environment Ministers agreed in 1992 a Directive to reduce the sulphur content of diesel and gas oil fuels to 0.2 per cent in 1994 and then of diesel to 0.05 per cent by October 1996. A proposal further to reduce the sulphur content of heating, industrial and marine gas oils to 0.1 per cent by 1999 was contested by the industry, and the Council adopted instead a proposal to review the next steps on sulphur during 1994. The sulphur problem cannot be limited to vehicle fuels and industrial use, since bunker fuel and aviation fuel are also an important source of global

---

[4] The latest idea is that subsidized surpluses of European agricultural alcohol should be exported to Brazil to make good the shortages in their biofuels programme (itself subsidized). This will increase the pressure on the Brazilian oil refining industry to export surplus gasoline, some of which will enter the European market, competing with unsubsidized European refiners.

sulphur emissions. These can be controlled at a European level only for inland waterways and perhaps commuter airlines.

Vessels and aircraft in international trade will generally have options to refuel elsewhere, at lower cost. CONCAWE (the Oil Companies' European Organization for Environmental and Health Protection) and the International Maritime Organisation (IMO) have both studied this issue. The IMO has examined an international limit of 1.5 per cent on bunker fuel. At this point the issue is one of relatively minor alleviation of Europe's sulphur emissions requiring very widespread international support and imposing very large costs of adaptation on the refining industry. The sulphur content of European refined products has already fallen by 60 per cent over the 1980s, according to a CONCAWE study. To achieve this there has been an increase in refining intensity and therefore in fuel consumption and CO_2 emissions.

Sulphur in industrial fuel One large load in our environmental caravan is the EC Directive 609 of 1988 on large combustion plants.[5] This sets national targets for reducing SO_X and NO_X emissions from plants with a capacity larger than 50MW. Virtually all refineries in the EU are, by definition, large combustion plants and subject to the same sulphur restrictions as the electricity generating sector and industry generally. According to CONCAWE data SO_2 emissions from European refineries were reduced by 40 per cent between 1979 and 1989.

Particulates Particulate emission from diesel fuel will become an increasing problem which may eventually disrupt the steady progress that diesel is making due to lower taxes at the expense of gasoline in many European markets. Diesel is giving more miles to the gallon. The Commission points out that its more extreme (and expensive) proposals for reducing sulphur in diesel fuel will also have a significant effect in reducing particulate emissions. The question is whether that is the only or the most cost-effective way.

---

[5] This is the directive which is forcing power stations progressively to fit flue gas desulphurization. It is also one of the reasons why gas has become so attractive as a fuel for new power stations.

Product quality effects on the oil industry The product quality story in Europe fits the 'caravan' image quite well. There are a number of issues, progressing at different speeds. The bad news is that many of them will cost a lot of money. At a very rough guess, the industry may already be spending $3–4 billion a year for an investment programme totalling around $20 billion which would not have been spent if there had been no new environmental regulation in Europe during the past five years. There might be another $50 billion programme along the road in the next ten years if all the proposals I have described were adopted, including the reductions in sulphur in diesel and fuel oil and related reductions in benzene content in gasoline and diesel. However, these estimates all assume the worst as regards timing, implementation and learning to reduce the cost of achieving acceptable results.

The good news is that a healthy degree of dialogue has developed between the Commission and various industry bodies, including EUROPIA (the European Petroleum Industry Association) and CONCAWE. The Commission in 1992 set up a tripartite consultative group involving itself, the automobile industry, and the petroleum industry. This will study technical interactions between fuel, engines, and air quality; develop a framework and models for analysis of environmental, technical and cost efficiency; and develop a framework for a general air quality directive. This is a new approach, but a promising one. This inter-industry approach should be possible to prevent the politics getting ahead of the science, as happened with the US Clean Air Act. It should also provide a spur to the scientists and to the industry to come up with timely and constructive solutions to the environmental problems.

In this and in other forums many companies share the objective of promoting discussion between the industries concerned and the environmental authorities, to seek as far as possible:

- Agreement on the criteria for solutions. The industry has a chance to demonstrate its argument that it is the result that matters: regulations should focus on what comes out of the tailpipe rather than what goes on inside the engine.

- Exposure and resolution of conflicts between objectives, for example between reducing CO_2 and increasing processing intensity to remove sulphur or VOCs.

- Identifying for public debate a range of solutions with their costs and impacts.

The lost markets EC Directive 609 of 1988 on large combustion plants affects oil industry customers as well as its own operations. The Directive sets national targets for reducing SO_x and NO_x emissions from plants with a capacity larger than 50MW. The main effect is on the demand for high-sulphur fuel oil and indirectly on the mix of the oil products demanded. Most important for the oil industry is the effect on the electric power sector. Natural gas is the fuel of choice for most of the new power stations planned or under construction in Europe. The main immediate driving factors are the sulphur directives on large combustion plants, the lower capital cost per unit and modular increments versus combined cycle generating plants versus conventional steam cycle plants and, not least, the fact that gas availability for the next ten or more years seems to ensure gas prices which will compete against construction of new oil stations at current and possibly rising prices. The deregulation of the industry in the UK also provides new openings for independent power producers and co-generators for whom combined cycle gas turbine plants are the natural entry vehicle.

The question of fuel switching is too large to pursue here. In the long run it must result in a widening of the differential between high-sulphur and low-sulphur crude oil. There are difficult questions as regards potential investors in desulphurization and residue destruction while those margins are uncertain in size and timing.

The free riders These discussions still do not involve directly the people responsible for a major part of the remaining pollution by the use of vehicles – their owners and drivers. *Sub-optimal driving practices and poor motor vehicle maintenance* are major contributors to pollution, mainly through old cars and trucks which remain on the road. These contribute pollution quite disproportionate to their number and use. Californian studies have shown that in their situation tonne for tonne reductions in emissions could be achieved through vehicle maintenance at one-eighth of the cost of the reductions expected from the next round of Clean Air Regulations, and similar tonnage reductions could be achieved at half of that lower cost by a programme to scrap older vehicles. Those estimates do not take account of the benefits of reducing risk of accidents due to faulty vehicles.

Transportation and oil spills: old problem, new urgency
Major pollution from accidents accounts for only a small part of the oil discharged into the seas worldwide (most of it comes from bad-practice operators washing out tanks or discharging ballast at sea) but it is the most obviously damaging to the environment because large volumes of oil may reach environmentally sensitive locations in high concentrations, under conditions where the containment and recovery of the oil is difficult to achieve. Spills can occur from production or drilling platforms but so far the industry has a good record. The main concern in Europe has been with spills from tankers.

In 1993 there were two accidents involving offshore oil spills in Europe: the Braer off the coast of Shetland and the smaller British Trent incident off the Belgian coast. I will not discuss the particulars of the cases here: the point is that they gave extra urgency to public and government concern about the risk of marine pollution which has been a major issue for the world industry since the Exxon Valdez spill in 1989.

As a result we are facing the need to provide a quantum jump in the environmental security of oil cargoes carried by sea. Cost and management implications for the industry could be severe. The evolving policies on marine transport of oil will have a profound effect on the shipping industry and the relationship between shipowners, shippers, and the major oil companies. There are diverse interests, since responsibility is divided between ship and terminal owners, operators, oil companies and authorities such as port authorities and coastguards. All must play a part in providing solutions which work.

International and multinational necessities Few solutions to maritime problems are within the reach of one country's regulatory authority or even of the European Union. However, as we saw in the US after the Exxon Valdez episode, events may force countries to take unilateral action even though the solutions are partial. In that case international competition is distorted, and there is a knock-on effect as disadvantaged parties try to get protection for their position. In looking for industry solutions, therefore, I believe we have to seek a mix of approaches by local and regional authorities, by the oil and shipping industries, and by governments through international agreement. There is no need to invent and negotiate an entirely new international regime. Ratification of existing IMO protocol and strict enforcement by national authorities would go a long way to change the situation.

The real point in both the short and long term is to spend more money on inspection, data analysis and enforcement. This must be universally applied if it is to have an effect. The major oil companies control less than one-quarter of the ships moving oil by sea, and are already, in general, suffering cost disadvantages through the application of standards in selection of vessels to charter. BP, for one, has indicated a willingness to support an EU levy on all ships discharging or lifting oil in EU ports, if the funds are used to provide a more comprehensive ship monitoring and inspection system, backed by real powers to detain severely substandard ships and refuse them port facilities.

Impact on the industry I have spent almost as much time discussing the transportation question as product quality. I make no apologies for this: most international oil trade is transported by sea. The more difficult and expensive such carriage becomes, the more difficult it will be to maintain the competitiveness of oil imports against domestic alternatives. Marine oil spills are a real risk and a risk which has immense public impact on the industry and the commodity which it trades. These risks must be reduced if we are to carry on our business in a reasonable relationship with the public on environmental issues generally. The environmental damage caused by the Exxon Valdez oil spill damaged the whole industry as well as Exxon. Ways must be found of establishing standards which will bind large numbers of international operators of very varied practices. Without that discipline the oil and shipping industries will not achieve the 'quantum jump' that I believe is expected in reducing the risk of major oil spills.

Operations and sites
Like every other industry, the petroleum industry's own operations are affected by various environmental policies. These affect current and future operations and responsibility for the consequences of past operations. Different countries have devised different plans to meet these targets. Some (like the sulphur restrictions referred to above) are simply applications to the oil industry of

restrictions which apply to industry generally. Others are more specific to the oil industry, and are driven by concerns about preserving ecosystems, rather than simply human health.

Waste now averages only around 0.1 per cent of refinery throughput. This figure will be reduced further. There have been even greater reductions in oil discharged as effluent. The costs and benefits of waste reduction programmes are quite complex: it so often happens that waste is reduced by tighter operating procedures which do cost money but also yield improved process control, reliability and other direct economic benefits.

Habitats In Europe, most prospective oil exploration and production areas are offshore and in relatively deep water. The industry does not face the 'habitat' problems of protection of wetlands and environmentally sensitive areas on the same scale as in North America. Its main environmental baggage upstream in Europe is the discharge of water into the North Sea. The industry is voluntarily moving to re-injection, combined with hydrocyclone treatment for disposed water. Oil from operations is limited by UK legislation to 40 parts per million and represents a very small environmental problem compared to the major pollutants of nitrogen and phosphorus from the run-off of agricultural fertilizers and heavy metals from the discharge of raw sewage.

The other big concern is the discharge of oil-based drilling muds and cuttings. Here the amount of oil discharged has been halved since the mid-1980s, due to a strategy of enhanced cleaning, recycling and re-injection. Oil-based muds are superior to water-based muds in the tough geological conditions of the North Sea, permitting longer horizontal reach for the wells and therefore fewer and lighter platforms than would be possible with the use of water-based muds.

Site liability There is a further major burden of environmental policy which is bringing a crowd of lawyers into the caravan. Nice people I am sure, but troubling to the industry in Europe. The question in every country is liability for cleaning up polluted sites or for site-related damages, such as groundwater pollution. Then there is the question of whether liability laws in different European countries should be 'harmonized', or developed in each country on the basis of 'subsidiarity'. The Council of Europe has proposed a new international (European) convention for environmental damage. The European Commission has published a 'Green Paper' – that means a consultative paper rather than a policy position – on 'Remedying environmental damage'. This is the obvious basis for discussion over 1994.

Legal issues The EC Commission's Green Paper identified the key issues for policy decision, whether on a national or a European basis. The first is the nature and extent of civil liabilities: whether they are to be *strict* liabilities, depending only on establishing a causal connection between the damage and the polluting activity, or to be *fault-based*, which would require proof that the 'polluter' committed an act of negligence, or omitted to carry out a duty of care, including duties prescribed by regulation and also duties which had arisen under more general civil law. The German Law on Environmental Liability of 1990 introduces strict liability for environmental damage caused by industrial operations. Elsewhere strict liability is generally applied to waste disposal and effluents. Separating environmental liabilities from other forms of civil liability can produce legal complexities which may do more for the lawyers' business than for the environment.

Clearly, the more legislation moves in the direction of strict liability, the more important it becomes to establish whether there are acceptable categories of defence: for example if the polluting activity was, at the time when it was caused, in compliance with existing pollution control laws and whether the then 'state of the art' technology was used to mitigate environmental damage. Some countries, such as the UK, have a strong legal tradition against retrospective legislation. Next, there may be difficulties in allocating 'fault' between the current owner and an historic polluter. Then there is the question of apportioning liability: is it *'several'* (in which case each of several parties bears liability according to his share of the responsibility)? Is it *'joint-and-several'* (in which case any of those responsible can be called upon to meet the whole liability, but has the right to seek recovery from the other parties involved)? Or is it *'joint'* (in which case those with deep pockets – or those most easily brought to court – can be liable for the whole damage and have no right to seek recovery from the other parties involved)? The EU principle of 'the polluter pays' goes against the concept of joint liability.

The worst case would be a combination of strict, joint and retrospective liability, putting pollution on a level of culpability which is difficult to apply even to war crimes.

Insurance The issue of liability reacts critically with the questions of the terms on which future insurance is available to the oil industry, and whether the insurance industry's exposure to claims arising from past pollution is adequately covered.

Funds and superfunds The questions regarding liability may induce 'good behaviour' by potential polluters. They do not necessarily address the problem of correcting damage where the polluter cannot be identified or is beyond reach. This raises the question of establishing public funds from which to compensate victims of environmental damage in cases where the liability of those currently connected with the site is limited, whether in civil law or by statute. In general, European countries do not have compensation funds available for use in such cases. The European Commission Green Paper raises the possibility of 'collective compensation' from the industry concerned in cases where a polluter cannot be identified – for example in the case of acid rain. This is also a kind of retrospective punishment in which payment is levied not on the polluter but on those who undertake the investments necessary to prevent pollution. The US Superfund is another discouraging example, in terms of the costs levied or the use of the funds raised. However, I suspect many in the industry would prefer to see a role for a public compensation fund to be developed if that was the only alternative to the arbitrary assignment of damage liability to parties whose connection with the pollution is remote or minimal.

Effect on the industry Uncertainties about liability have several effects on the downstream business. They increase the risk of being in the business. They also make it much more difficult to get out of the business or even to get out of sites that are redundant in today's market. The costs of closing down are high anyway, but uncertainties about how much cleaning up must be done and what liabilities remain make it difficult to dispose of sites. High and uncertain exit costs mean that facilities are kept going against low cash returns.

Funds for more positive investments in continuing operations are restricted, and the necessary restructuring of the industry is slowed down. The caravan would move faster if it were easier for those who dislike the course to drop out.

Carbon dioxide and climate change

What kind of a problem is it?

CO_2 emissions are in a quite different category from all the other environmental policies I have discussed. This monster is really not even travelling in the same caravan: it has hardly started the journey, there are some crusading slogans up there, a lot of advisers buzzing around, and a great deal of reciprocal scepticism between the believers and unbelievers.

First, CO_2 is not a proven pollutant in the same sense as lead and sulphur. As already indicated above, the concern about CO_2 arises solely from suspicion about its possible contribution to global warming.

The policy arguments which flow from this concern are familiar. Briefly stated, the argument is that the risk of climate change is sufficiently serious to justify 'precautionary' policies to reduce the rate of CO_2 emissions. Targets have been accepted by European national governments to reduce the rate of CO_2 emissions in Europe to 1990 levels, or some fraction thereof, by the year 2000 at the earliest. Some other countries have made similar commitments following the Rio Convention. National plans to fulfil this commitment are being drawn up. These plans will be reviewed at a conference of the parties to the United Nations Climate Convention, that first meets in April 1995.

It is extremely unlikely that by that date the scientific understanding of global warming and estimates of its probable climatic effect will have become much more precise than they are today – if anything, I suspect that the sense of uncertainty will be heightening. In that case, there will be a tendency to look for more, not less, precautionary policy. On the other hand, governments may realize, as they work through their national plans, that stabilizing the rate of CO_2 emissions is going to be difficult and expensive; that stabilizing and holding the 1990 rate for ever will cost more and more as the pressure of economic growth demands more and more energy; and that the ability of the OECD countries alone to influence the global level of CO_2 concentration is diminishing fast as economic growth proceeds rapidly in East Asia, China and other developing countries. The problem will not have gone away and it will seem more intractable.

Secondly, 95 per cent of the CO_2 emitted into the atmosphere by oil combustion is produced by consumers, not by the oil industry. Once customers have switched fuels to the extent that is practical – which in Europe in the short term tends to mean switching to natural gas in power generation – there is nothing more the supply side of the industry can do about CO_2: reduction must be made by consumers using less fuel.

Thirdly, there is generally a conflict between the environmental policies to improve the environmental quality of petroleum products, and policies to restrain CO_2 emissions. The simple reason for this is that over the past decade the increased efficiency of over 20 per cent in the use of energy in refineries – and therefore reduction in CO_2 emitted by burning refinery fuel – has been more than offset by the increased processing required to produce lead-free gasoline of an acceptable octane, to reduce the sulphur content of gas oil, and to increase the conversion of residual fuel to lighter products as heavy fuel oil has been substituted by lighter fuels, natural gas, and nuclear electricity. The EU proposals for further reductions in sulphur will continue this process.

Where the politics are

On the political front the idea of carbon and energy taxes in Europe and energy or BTU taxes (British thermal unit) in the US has not fared well. The political experience has been that, contrary to expectations, the public has not been more willing to pay 'green taxes' than other taxes. The carbon/ energy tax is still an idea twisting in the Brussels wind. The BTU tax (at the time of writing) has disappeared from the US legislative agenda. In the UK, the application of the 17.5 per cent 'value added tax' (VAT) rate to electricity and natural gas has gone ahead, despite immense unpopularity, thanks to the unique party discipline normally available to the British government in its parliament. The public attitude to taxes does not always follow economic logic. For example, the extension of VAT in the UK, far from being a discriminatory tax, actually levels the 'playing field' between the fuels concerned, which were previously exempt, and all other goods, including energy-efficient appliances which are subject to VAT.

The future of ecotaxes

This political experience may cool the idea of carbon taxes somewhat, but it is not dead. It was always part of the carbon and ecotax arguments that they could serve a double purpose, the so-called 'double dividend':

- First, ecotaxes would induce consumers to modify their behaviour in an acceptable way, i.e. by reducing the use of fossil fuels in proportion to their carbon:energy ratio.

- Second, the revenue raised by ecotaxes could be recycled to reduce taxes which are economically damaging in other ways, for example by reducing payroll taxes on employment and the double taxing of dividends on capital.

These are still good arguments – arguments for including fuel excise taxes in any general plans for tax reform. The same economic logic suggests that existing fuel excise taxes should be reformed to reflect environmental objectives. In plain language, this generally means increasing taxes and reducing subsidies on coal and probably reducing taxes on oil, as was done in Sweden when a carbon tax was introduced there. But that is not the argument we hear. The latest rumour, based again on the dubious assumption that people are more willing to pay ecotaxes than other taxes, is that carbon and other taxes should be raised to cover the increased burden of the social costs which in most EU countries are imposed by legislation loosely identified by the phrase 'Social Charter'. It is impossible to tell what the public reaction would be to any scheme of tax reform, which raised significantly more revenue from energy or environmental taxes in order to reduce other taxes. Who wants to start opening the subject up in this way?

New trends in policy

Outside the political arena, there has been some development in the intellectual debate about taxes, regulation, voluntary agreements and their respective roles. The OECD Secretariat organized a conference in Paris in June 1993 to review a range of studies dealing with the impact of carbon taxes, methods of estimating the benefits of mitigating climate change, and the merits of various policy instruments. The conference was very valuable in emphasizing the tools which economics can bring to this debate. Unfortunately, and I speak as an economist, the tools are far from adequate to deal with processes which are so little understood and phenomena about which there is so little data – which is true of the future generally. My impression from this meeting was that there is increasing intellectual acceptance of the idea that carbon taxes or tradeable permits are not the only policy instruments which can or should be used by countries in implementing their national plans for CO_2 reduction. Much was said about the merits of a 'mixed bag' of regulation, taxes, permits, and voluntary agreements. I even heard economists say that the tax rates quoted in some of the modelling exercises should be regarded simply as 'indexes' of the sets of policies necessary to achieve the estimated results.

Dynamics of business versus policy
This new policy approach of the 'mixed bag' – if it is new and if it is an approach – contains both good and bad news. The good news is that it recognizes that there are situations in which other measures can be better than taxes. I have personally long argued that taxes which are imposed on consumers before there are alternatives available to them are punitive and unstable. They are punitive because – in the case of transportation – if there has been no prior investment in more efficient vehicles or in public transport the consumer has no choice but to pay more for his/her unavoidable travel. The elasticity of demand of transport fuels in the short run is very low, as we all know. The taxes are unstable because as investments are made in alternatives and choices become available to the consumer, he/she can respond to higher prices by different behaviour – his/her elasticity of demand increases. This means that a lower tax rate is required to achieve any target reduction in the long run than in the short run. It is risky for suppliers of alternatives to bet on markets dependent on all the normal commercial and economic uncertainties plus uncertainties about the tax rate which is bound to be unstable for the reason I have just given.

These difficulties might be eased if taxes were coordinated with other policies – public investment in those transport infrastructures which lie within the public sector, and the investment plans of suppliers of more efficient vehicles (for example). In market economies, of course, there is no mechanism for coordinating the investment plans of the private sector except taxes and regulations.

In these circumstances, regulations such as the CAFE (Corporate Average Fuel Economy) regulations can play a risk-reducing role.[6] If all manufacturers know that by a certain date a proportion of their products must meet a certain efficiency standard, or all builders know that by a certain date all their building must incorporate certain energy saving designs, then the competitive risk is simply how to achieve that result most cheaply and attractively for the final purchaser. The history of other environmental policies illustrates this approach: there have been tax incentives to use unleaded fuel during a transitional period but there have been regulations to ensure that by certain dates all cars sold will be capable of running on unleaded fuel. The key point is to achieve coordination between taxes and incentives to develop alternatives, and to avoid conflicting signals.

Industry impact

The bad news is that a 'mixed bag' of policy instruments can also conceal policies which have little to do with the environment and a lot to do with raising money. There is a danger that the consumer, and the industries supplying the consumer, can get the worst of both: badly designed regulations and heavy taxes.

Certainly the experience of 1993 has been that vague talk of carbon and energy taxes has produced increased gasoline taxes which are not large enough to induce significant alteration in transport demand, and have no effect on the competitive position of low-carbon fuels such as gas and oil versus high-carbon fuels such as coal, but transfer a lot of money from motorists to government. This is not a good precedent for the petroleum producers. I am amazed that in the US debate the American Petroleum Institute – or at least

---

[6] There are particular risks in a competitive industry (such as the automobile industry) in being either first or last to invest in bringing to market a new model – especially something as different, for example, as a hybrid electric car. As some manufacturers found after 1973, one can invest too early, and wait for an undefined length of time with unsaleable products. Alternatively, one can wait for the trend to change and then invest to catch up – generally a better defined risk. This simple business logic, I think, explains why every major automobile manufacturer has an electric or hybrid car on the drawing board and carries out research to keep the design at the state of the art – and probably update the production plan – but never actually produces except in experimental quantities. The same story holds across a wide range of the energy efficient equipment which really serious carbon taxes might eventually bring in, from 'long-life' light bulbs to cladding buildings with photovoltaic cells.

some of its members – regarded it as a triumph that no precedent was created for taxing fuels which compete with oil and the revenue burden was eventually loaded entirely onto gasoline.

For Europe, however the bag of CO_2 policies is eventually mixed, the one certain winner is natural gas. Lower CO_2 emissions per kW generated are an added bonus, reinforcing trends driven by other policies such as sulphur reduction.

The consumer confidence trick
At a rough estimate, OECD governments in Europe in 1991 collected $120 billion in consumption taxes on petroleum products, excluding VAT (which applies widely outside the oil sector as well).[7]

These payments, spread over all oil consumption in Europe, averaged $24 per barrel, or approximately $200 per tonne of carbon. In the theory of environmental taxes, very large environmental and climate costs have been 'internalized' in Europe. In the context of the global warming debate, the real misfortune for Europe is that governments have accepted targets for CO_2 emissions based on reductions from 1990 levels. The 1990 levels were the result of a history of high consumption taxes, and related high prices, compared with the low tax OECD countries such as the US and Canada. No credit has been gained for the lower levels of emission per head or per unit of GDP which higher taxes have generated over the years in Europe, compared with North America. The quantitative target reductions in the UN Climate Convention mean that European consumers will be asked to pay twice for the alleged 'external' costs of burning fossil fuels. The political reality is that it is gasoline consumers who will pay, while the coal industries of Germany and the UK continue to enjoy 'negative taxation' in the form of subsidies and protected markets.[8]

---

[7] A further $4 billion was collected by the European OECD governments, mainly the UK and Norway, in royalties and special taxes on petroleum production. Governments of countries exporting oil to Europe probably collected another $40 billion in 'take', including taxes, profits, and returns to state investments: few exporting governments publish accounts which enable these figures to be reported precisely. The division of rent between producer and consumer governments is not an environmental question and I do not discuss it here.

[8] It is interesting that most of the 'Green' movements in the UK did not campaign for the unpopular cause of reducing these subsidies, and thus closing down British coal mines, when the British government needed support on this issue towards the end of 1992.

The deep paradox of oil

So we are left with the fact that, so long as the oil market remains in surplus, consumer governments can maintain and sometimes increase their share of the economic rent which the merits of oil provides. As we argue about the share we miss the main message about the size of the rent which is available for tax. Consumers pay these taxes, without much complaint, because oil enables people to move about in vehicles for which there are no effective competitors, and carry out activities which without oil it would be difficult to afford, or even imagine. The size of the consumer tax take is a solid measure of the economic value oil has for consumers.

Moreover, oil is not only economically more valuable, but for a whole range of activities, using oil produces a more acceptable environmental impact than many available alternatives. Of course, natural gas is a cleaner fuel which is cost competitive for some purposes. But if gas were to replace oil everywhere today, the world's proven gas reserves would be exhausted in ten years and oil would be back. Nuclear does seem to be regarded as an acceptable alternative and coal is certainly inferior to oil by a variety of environmental, health and safety criteria.

Summary: where are the companies going?

Where does all this leave the oil companies and what are they doing about it? There are, of course, differences between the companies and I will try to make general remarks rather than try to speak for all. There will be someone who disagrees with almost every single point I make.

First, complying with changing environmental policies is now a major task of management at all levels. The caravan is on the move and the oil industry is travelling with it whether it wishes to or not. The industry cannot stand still. Many companies build environmental objectives into general management objectives, measure performance against them, and reward accordingly. Environmental matters, like safety and health, are too important to be left to specialists (though the specialists are needed, more than ever before).

Secondly, it is not enough to behave well by objective criteria. The industry has also to be seen to behave well by the subjective criteria of opinion-makers, many of whom are very sceptical about the motives and manners of 'big business'. There are three keys to this approach:

- Reveal more information, beyond the disclosures required by law both in content and in the degree of publicity the industry gives to the disclosures.

- Build practical working relationships between the operating site and the community on environmental matters. Local pollution – or the suspicion or perception of it – is the driving force for support for more general and less objective environmentalism.

- Participate in the policy debate in a reasonable way. The industry should try to demonstrate to the European Commission, and to the member governments, and more importantly, to the public behind them:

  (a) That the industry does care about environmental objectives.
  (b) That economics do matter: choices have to be made regarding the allocation of scarce resources.
  (c) That technology can get better environmental results and reduce the cost, sometimes in ways we cannot completely foresee.

In Europe there seems so far to be room for such an approach, but on almost every count the oil industry cannot act alone. It needs the automobile industry, the shipping industry, the insurance industry, and not least the consumers and governments. I have mentioned the joint oil, automobile industry, and Commission study on fuel quality, the Commission Green Paper on liability for environmental damage, and the ongoing consultations about safe transport of oil at sea. The bad news is, of course, that, however cost-effective the environmental result is, there will still be costs. These will be difficult to recover in refining and retailing margins, because most of them will be capital costs. The question is whether these additional competitive pressures will force some restructuring in the industry.

Thirdly, global warming is different. Product quality, liability, and transport issues are common to many industries. Global warming is a special problem confined to fossil fuel use. Consumers, not producers, will bear the cost of change first. If climate policies eventually require less energy to be used, our customers must adapt their industrial, transportation, and consumption apparatus and habits to use less energy. The effect on the energy supply industries will come later, through eventual changes in the balance of supply and demand for fossil fuels and the sharing of the fossil fuel market between fuels of different carbon intensity.

I stress the word 'eventually'. Future climate policies are much less certain than the policies connected with air quality, sulphur, and oil in the North Sea. The benefits of mitigating CO_2 emissions are remote in time, unknown in scale, and uncertain in location. The world is at a very early stage in estimating the costs of policies to reduce CO_2 emissions. The caravan of global climate policies is still assembling. It may not be moving very far in any credible direction, but do not look away: this caravan is not going to break up soon.

For the time being we will see a mixed bag – perhaps a ragbag – of policies on CO_2 being spread on the ground for the policy-traders to pick over. I doubt if the politics will permit the main weight of those policies to fall on the fuel which emits the most CO_2 per unit of energy used – that is, on coal. In reality, as we have seen 1993, ecotaxes will tend to mean gasoline taxes. This ought to give the oil industry the opportunity to raise questions about the structure of fuel taxation: current excise taxes bear no relation to CO_2 problems – or even to sulphur problems.

Who will speak for oil?

This leads to the larger question of why the economic and environmental merits of oil are not better recognized and not more widely promoted. There is only limited scope for cheaper and cleaner energy sources than oil. Without oil the world would be a poorer, dirtier, and eventually perhaps a warmer place. Who can put this case? The industry is divided between companies and governments, exporters and importers, traders, refiners, retailers and producers. Each player in the industry carries a particular piece of environmental baggage and is trying to manage its load along with other burdens. This very natural state of affairs may, however, mean that the caravan moves away from an oily path to a route which will be worse for the consuming public and the population at large in terms of cost and environmental result. It is very difficult for any one segment of the industry to do much. The companies have limited resources and relatively narrow commercial interests. Governments of exporting countries, whether OPEC members or not, may also appear partisan for other reasons. The consumers are not organized except through the market as customers. I am not going to venture here to answer the question 'Who will speak for oil?' but I suggest that the problem is one which needs to be considered carefully by the industry and those connected with it either as customers or as related enterprises.

Politics, Economics, and Environment: Experience of the US Oil and Gas Industries

Clement B. Malin

Abstract

Creating and maintaining a safe, clean environment is an important part of the worldwide effort to improve living standards and the quality of life for the human race. And, as world population grows and more nations industrialize, it will become increasingly difficult, but no less important, to balance environmental protection with economic development and the provision of affordable energy, to achieve sustainable growth.

Over the past 20 years, the United States has been the focal point for the debate, exploration and, in some cases, resolution of these issues. And, as providers of much of society's energy, the US oil industry has been a major participant in this process. Now, as the debate gathers momentum in other countries, it is appropriate to review what has transpired in the US, to learn which courses of action have worked, and which have not.

Background

The challenge the oil industry faces is how to provide the affordable energy that society needs for economic growth and improving living standards, while protecting the environment. We have made substantial progress towards meeting that challenge, producing cleaner-burning fuels, reducing exhaust emissions, and improving the efficiency, cleanliness and safety of our vast operations. This has been done not without significant investment and increases in operating costs, not always recovered in the market place. Moreover, in the United States, the oil industry continues to operate under laws and regulations that often are costly, inconsistent, contradictory, complex and burdensome. In concert with an explosion of litigation, these policies combine to burden US industry, taxpayers and consumers with unnecessary costs, sap the nation's competitiveness, and drain away needed investment capital, entailing the loss of thousands of jobs.

The role of government

A great many of these laws concern environmental protection. Regulation in the US costs the economy an estimated $400 billion to $500 billion a year. This estimate does not include the cost of compliance – as yet undetermined – with the Clean Air Act of 1990 or potential amendments to Superfund legislation.

Economists generally agree that, to reduce the rate of unemployment, real GDP must increase by about 3 per cent a year. However, according to the US Environmental Protection Agency, without legislative and regulatory reform, total yearly costs for all pollution control activities in the US could reach 2.8 per cent of GDP by the year 2000.[1] This substantially increases the GDP growth required level to achieve unemployment reduction, thus adding a real burden to US job creation in the form of cumulative environmental legislation.

Many environmental, health and safety laws, such as the US acid rain legislation, have been enacted without first determining whether the risk addressed is sufficient to warrant the remedy imposed. These laws often have

---

[1] Environmental Protection Agency, *Environmental Investments: The Cost of a Clean Environment*, EPA, Washington DC, 1990.

unintended, costly and counterproductive results, and divert funds from addressing other societal needs.

This practice has been encouraged by some environmentalists and policy-makers, who believe that economic costs should not even be considered in drawing up legislation regulating environmental and health issues. In fact, US courts have ruled that the Clean Air Act does not require regulators to apply cost-benefit analysis to the establishment of ambient air quality standards.

To add insult to injury, US environmental laws often conflict with each other, creating confusion within industry and among investors, as companies attempt to respond to them. Since 1970, the US has passed a series of laws and regulations concerning the oil industry that send mixed signals. On the one hand, the US government has mandated massive investment by the oil industry to produce reformulated gasolines. On the other, it has taken steps to diminish gasoline demand by as much as 30 per cent in 20 years, through alternative fuel programmes. And recently, the administration announced a joint industry–government project to develop automobiles that use far less fuel – and may not run on oil-based fuels at all. Such contradictory policies increase investment risks for the oil industry, since we must commit billions of shareholders' dollars to capital spending projects, to serve the public most efficiently, only to face the possibility not only of being unable to recover the costs of these investments, but also of having capital equipment rendered obsolete well before the end of its useful economic life.

Examples of legislation

There are a number of examples of legislation in the United States where initial legitimate aims driving the demand for legislation were undermined in implementation, where excessive regulation resulted in unnecessary invest-ment, and where solutions were undertaken without an understanding of the problem.[2]

---

[2] See also Valerie Fogleman's paper above, 'Economic Impacts of Environmental Law: the US Experience and its International Relevance'.

Superfund

Superfund, officially called the Comprehensive Environmental Response, Compensation and Liability Act, was passed in 1980 to 1) clean up old hazardous waste sites, 2) respond to waste emergencies, and 3) provide incentives for sound waste management. The intention of the law was to make those parties responsible for an environmental problem pay for its remedy (i.e. the 'polluter pays' principle). To deal with cases where the source of pollution could not be traced to a responsible party, Congress established a trust fund of $1.6 billion to oversee clean-up. This is to be paid for by an excise tax on petroleum and chemicals, an environmental tax on corporations, a general revenue appropriation, and a motor fuels tax to fund the EPA underground storage tank fund.

In 1986, Congress increased the trust fund by $8.5 billion, to address specific clean-up goals and standards, and to establish long-term solutions and effective response mechanisms. (The US oil industry currently pays between 50 and 60 per cent of this cost, even though it is responsible for no more than an estimated 10 per cent of the wastes. The proportional liability of the oil industry was mandated by Congress through mechanisms such as the Superfund import fee, without a corresponding determination of actual physical liability. So much for 'polluter pays'. Texaco alone paid an estimated $100 million for 1989–91.)

Total projected costs for clean-ups nationwide are now estimated at between $500 billion and $1 trillion, by the time all planned work is completed. However, so far, about 85 per cent of money spent has been for lawyers' fees and other administrative transaction costs.

One reason why Superfund costs have grown far beyond expectations is the extremes to which its provisions are often carried. At one 81-acre former industrial site in Mississippi, soil was determined to contain two ounces of hazardous materials per ton. The EPA ordered a complete clean-up of the site. Some experts concluded that this small amount of contamination was harmless, and proposed spreading a layer of clean topsoil over the contaminated soil at a cost of $1 million. Instead, the EPA ordered Reichhold Chemical, the former occupant, to dig up more than 12,500 tons of soil and transport it to a commercial landfill in Louisiana. Estimated cost: $4 million. EPA's reason: should houses ever be built on the site, the soil must be safe enough for a child to eat half a teaspoon of dirt every month for 70 years and not contract cancer. The EPA has since acknowledged that at least half the $15 billion spent by the US to date on Superfund clean-ups was to comply with similar 'dirt eating' rules. Is this really a worthwhile deployment of taxpayers' funds?

The Clean Air Act

The Clean Air Act, passed by Congress in 1970, was the most comprehensive environmental legislation enacted to that time in the US. For the first time, it established national, rather than regional, air quality standards and deadlines for compliance. Thus far, it has improved air quality significantly, substantially reducing emissions of sulphur oxides, volatile organic compounds, carbon monoxide, particulates and, especially, lead. Indeed, organic tailpipe emissions from automobiles have been reduced by up to 96 per cent from 1970 levels.

This progress was also due to the improvements made by the US oil refining industry. Over the last three decades, its products have evolved, and differ markedly from how they were in the 1960s. Los Angeles had begun regulating gasoline olefin content in 1959. Significant use of oxygenates in gasoline for octane, volume or emissions reduction began in the 1980s. Their use was mandated in 1990, and oxygenates are expected to comprise 4–7 per cent of US gasoline by volume in the year 2000. Very little lead is used now, and it will be completely eliminated by 1996. Further, gasoline quality has been further enhanced with additive packages.

Despite the progress, however, some American cities still have not attained the national standard for ozone, carbon monoxide and particulate matter. In 1989, the US oil and automobile industries initiated research into ways to improve automobile engines and fuels, to reduce emissions and increase efficiency. They established the Auto/Oil Air Quality Improvement Research Program, drawing upon the companies' expertise plus that of outside research firms. However, before the results of this comprehensive effort were available, Congress passed sweeping amendments to the Clean Air Act, parts of which were based not on scientific data but on political expediency.

The Clean Air Act Amendments of 1990

The Clean Air Act Amendments of 1990 have a broad impact on industrial and manufacturing activity throughout the US, especially on the transportation sector. The legislation directed the EPA to establish regulations setting specific requirements for reformulated transportation fuels. Its goal: to reduce pollution by 56 billion lbs a year. Starting in 1994, motor vehicle tailpipe emissions will be further reduced. In 1995, reformulated gasolines with fewer aromatics will be required in cities with the worst ozone problems. Other cities have the option to participate and adopt the new fuel requirement, even if it is not necessary to meet federal standards (the 'opt-in' option).

Southern California has particularly severe ozone problems. To meet this specific need, that state will introduce in 1996 a pilot programme with even stricter emission standards, which apply to both cars and their fuels. Even though this programme was designed for California, the federal legislation allows other states to adopt it. States in the northeast have established a regional coalition called the Northeast States for Coordinated Air Use Management (NESCAUM) – a state government coalition – to address air quality problems there. Even though the California programme is not designed for conditions elsewhere, some members of NESCAUM favour adopting it; fortunately, others do not. This and similar efforts elsewhere raise the danger of creating an inefficient patchwork of federal, regional or local fuel standards that could lead to costly changes in the nation's fuel production and distribution system.

However, this expensive effort is aimed at only a small part of the pollution target. (See Figure 1.) If you were hunting an elephant, you would not aim at just the tail, to bring down the beast. But as this chart of hydrocarbon emissions for the state of Massachusetts indicates, that is what some environmentalists are advocating. By 1995, as the elephant shows, only 15 per cent of total hydrocarbons in the air will come from mobile sources, such as cars and trucks. The other 85 per cent will come from stationary and biogenic sources, such as power plants, fast-food establishments, dry cleaners and the natural decomposition of plants and trees.

There is also the question of diminishing returns for regulatory expenditures. An automobile produced in the US in 1990 was 96 per cent cleaner, in terms of organic emissions, than a typical late 1960s vintage model. The

Figure 1 Massachusetts 1995 hydrocarbon inventory

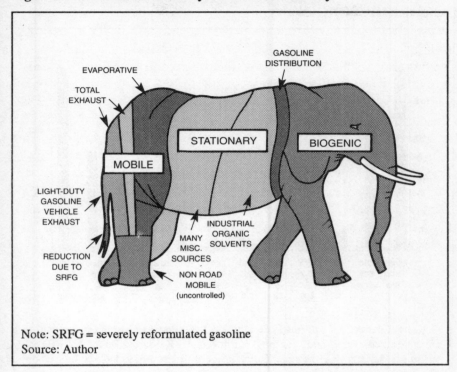

Note: SRFG = severely reformulated gasoline
Source: Author

estimated cost of this improvement: $1,200. In contrast, the standards mandated under the Clean Air Act Amendments will add a further reduction of only 2.4 per cent (of the 1970 model emissions), at a cost of $500 per car. The $500 is the cost of mechanical changes to cars, plus the cost of reformulated gasoline, which the Clean Air Act Amendments require, to reduce emissions. Depending on which reformulation is required by which states, the additional cost of manufacturing the gasoline could be anywhere from 5 cents to 25 cents a gallon. This amount may seem trivial, but we need to ask what benefit is gained, and whether some other beneficial environment objective could be achieved if the money the consumers are willing to pay was used in a different way.

These diminishing returns are best illustrated by a chart of mobile controls, based on a study in California. (See Figure 2.) It shows the various costs of

Figure 2 California incremental cost effectiveness vs. emissions reduction (HC or HC+NO$_x$)*

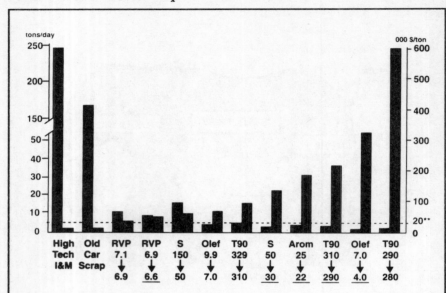

Notes: * Relative to California Federal Phase 1 (pertinent California specifications are underlined); ** $20,000 per ton is California Air Resources Board (CARB) criterion for determining the practicality of a control strategy; HC=Hydrocarbon; NO$_x$=Nitrous oxide; I&M=Inspection and maintenance; RVP=Reid vapour pressure; S=Sulphur; Olef=Olefins; T90= a term which refers to separating the heavy fraction of gasoline; Arom=Aromatics.
Source: Author.

removing or preventing the emission into the air of one ton of air pollution from an automobile. In each of the 12 pairs of bars, those on the right indicate cost – how many thousands of dollars it would cost to remove a ton of pollutants by various methods. Those on the left indicate effectiveness – the tons per day which each method would prevent or remove from the air, if fully and properly implemented. And the dotted line indicates that $20,000 per ton might be a reasonable upper limit for the cost of hydrocarbon and nitrous oxide controls.

The pairs of bars on the right are the least cost-effective methods. Most costly of all is lowering the T90, separating the heavy fraction of gasoline.

Lowering it to 280 degrees Fahrenheit is one part of the radical reformulation that may be needed to produce the California gasoline. That would be most expensive of all, costing roughly $600,000 to prevent the emission of one ton of pollutants into the air.[3]

In contrast, at the other end of the chart, there are far less costly programmes. The most effective of all are high-tech inspection and maintenance programmes, which can prevent the emission of a ton of pollutants for $1,400.

The story of diminishing returns is similar for pollution from industrial plants, including the oil industry's refineries. Figure 3 covers both mobile and stationary sources of pollution, and controls on both hydrocarbons and nitrogen oxides. It shows the cost-effectiveness of various control methods in a variety of industries – the oil industry, automobiles, utilities and industrial sources. These are indicated by O, A, U and I on the chart. There also are steps society can take to control pollution, identified by an S. The amount of results for the money expended varies enormously. Should not least cost alternatives be undertaken first?

A recent report by the National Petroleum Council, an advisory body to the US Secretary of Energy, estimated that the cost of pollution controls and health and safety programmes in the US oil refining sector between 1991 and 2010 will exceed $150 billion.[4] This estimate would be at a rate more than double that of industry environmental spending in the second half of the 1980s. It includes capital investments and operating and maintenance costs of physical plant, but not the costs of producing cleaner burning fuels or of possible changes in the industry's storage and distribution system. Further, the report estimates that projected US energy industry capital expenditures for pollution control to the end of the year 2000 will exceed the net asset value of the total industry. (See Figure 4.)

To minimize these projected costs, the report recommended:

---

[3] The $600,000 figure is, to a large extent, a one-time capital investment. There are also recurring operating costs which will remain in place after the initial capital investment. The figure is intended to show the range of costs for the different options which are available to reduce tailpipe emissions. As the automobile fleet is gradually replaced and newer lower polluting cars replace older, high emissions cars, the higher capital cost investments have less comparative value.

[4] National Petroleum Council, *US Petroleum Refining, Meeting Requirements for Cleaner Fuels and Refineries*, NPC, Washington DC, 1993.

Figure 3 Hydrocarbon or NO$_x$ emission reduction, \$/ton

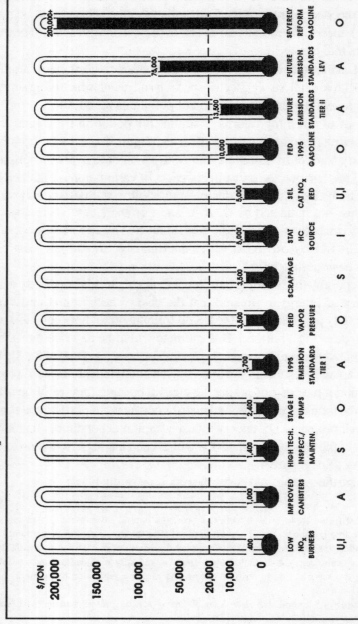

Notes: U=Utilities; I=Industries; A=Automotive; S=Society; O=Oil industry; STAT HC Source=Stationary hydrocarbon source; SEL CAT NO$_x$ RED=Selected catalytic nitrous oxide reduction; FED=Federal; LEV=Low emission vehicle.

Source: Author.

Figure 4 US refining industry projected capital expenditure vs. assets to the end of the year 2000

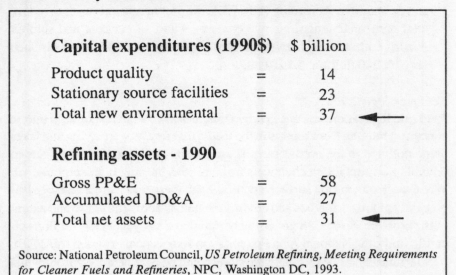

Capital expenditures (1990$) $ billion

Product quality	=	14
Stationary source facilities	=	23
Total new environmental	=	37 ←

Refining assets - 1990

Gross PP&E	=	58
Accumulated DD&A	=	27
Total net assets	=	31 ←

Source: National Petroleum Council, *US Petroleum Refining, Meeting Requirements for Cleaner Fuels and Refineries*, NPC, Washington DC, 1993.

- That the reformulated gasoline regulations that are implemented be fully compatible with the existing product distribution system.

- That policymakers understand that the cost of regulation will eventually be reflected in the marketplace (recent experience notwithstanding), and will affect the health and competitiveness of the US oil industry. The NPC study stresses the need for a measured, reasonable approach to environmental legislation, to preserve the balance between US energy industry competitiveness and the cost of environmental regulation.

- That a constructive partnership process be established among members of industry, government and other concerned parties, to find cost-effective solutions to public issues affecting the oil industry. An example of such a partnership process is the EPA's ultra-low sulphur diesel fuel regulation, enacted on 1 October 1993. Involving government, the oil industry, diesel engine manufacturers and consumers, it developed sound, cost-effective approaches to the issue. The government allowed adequate time for compliance, and the result was a level of diesel fuel desulphurization and engine modification that is cost effective and beneficial to the environment and the industry. According to the EPA, the programme promises to reduce total particulate emissions by between 19 and 38 per cent and sulphur dioxide emissions by between 43 and 80 per cent by 1995 at an estimated cost of $380 million to $910 million.

Acid rain legislation

Acid rain legislation was enacted by Congress in 1990 in response to public concern stemming from reports in the media that forests were dying and lakes being polluted in the northeastern US and parts of Canada, possibly due to precipitation containing chemicals from utilities burning high-sulphur coal. When the problem first surfaced in the 1970s, there was a great deal of disagreement among scientists and environmentalists about the cause and extent of the problem. So the US government initiated a ten-year, $500 million scientific study, the National Acid Precipitation Assessment Program (NAPAP).

Unfortunately, in 1990, before the study could be completed, environmental activists secured acid rain control measures in legislation reauthorizing the Clean Air Act. These measures sought to reduce utilities' emissions of sulphur dioxide by 10 million tons by the year 2000. They would do so by requiring many utilities to build scrubbers or more extensive emission reduction systems. Projected annual cost: $4 billion to $7 billion.

Between the introduction of the legislation and its enactment in 1990, the NAPAP report was released. It found that significant acid rain damage was limited to a relatively small area encompassing some 200 lakes and a few species of trees. The report recommended, as a cost-effective means of mitigating the effects of acid rain, that limestone should simply be spread over the affected areas, at a projected cost of $4 million, or one-thousandth of the cost of controlling the emissions of utilities. Nevertheless, environmental groups waged a campaign to discredit the study as politically motivated. They were successful, and Congress enacted the measures costing 1,000 times more than those suggested by the study. The unnecessary costs will be found in future electricity bills of consumers.

Regulations governing wetlands use

Regulations governing wetlands use, precipitated by the worthwhile goal of preserving environmentally valuable US wetlands, have been distorted into roadblocks that are being used to halt or delay projects, including oil and gas exploration and production, which do not threaten the wetlands. In 1989, the Army Corps of Engineers and the EPA issued a new wetlands delineation manual. The manual was developed without following standard US government administrative procedures. It initiated broad coverage and restrictions on vast areas of the US. Wetlands were defined, for example, as anywhere plants were found which were capable of surviving in conventional wetlands. Under its extreme applications, arid areas of Arizona could be classified as wetlands.

The US oil and gas industry was affected, since much of the nation's oil and gas resources are located in areas that are predominantly wetlands, such as Louisiana and Alaska. Wetlands regulations were turned into obstacles that significantly increased costs through delays, permit requirements, demands to avoid certain areas, and requirements to minimize and mitigate wetlands disturbances. As a result, the regulations limited and, at times, prevented significant energy projects, even though the disturbance to the wetlands would have been small. Currently, changes under consideration may correct these inequities. Meanwhile, however, the regulations are being used in a way that increases the cost of energy and other uses of that land – costs which are eventually passed on to the consumer.

Oil and gas exploration and production on the US Outer Continental Shelf

Federal policies dealing with this have been generally negative over the past decade. This has caused the industry to re-evaluate its investments in the US, and shift that money to projects outside the country. This costs American jobs; it is partly responsible for the fact that the US oil industry has lost 450,000 jobs over the past decade. It also deprives the federal government of significant oil and gas revenues, and unnecessarily increases US dependence on oil imports.

The Outer Continental Shelf is an enormous oil and gas producing zone. And its unexplored areas are projected to contain the majority of the remaining undiscovered oil and gas reserves in the US. Notwithstanding the fact that the Department of Energy, parts of the Interior Department, and authorizing legislation in Congress have been generally supportive of development, opposition on environmental grounds has come from other parts of Interior, the EPA, and sponsors of appropriations legislation in Congress. Moreover, some geographical regions support development; others oppose it.

The most discouraging examples of these inconsistencies come from areas where oil companies paid to acquire leases, conducted seismic analyses, and made a variety of other pre-development expenditures off southern Florida, North Carolina and Bristol Bay in Alaska. Due to local concern and environmental opposition, permits were withheld, moratoria imposed, and development stopped in these prospects for many years.

In the interim, the federal government has refused to refund the companies' investments. The US Office of Management and Budget has estimated the government's liability to the companies at some $1.5 billion, and has opposed compensating the oil companies because of the impact on the federal budget. Incredibly, companies are now forced to sue their own government. In addition to shifting investment away from the US, some companies, before risking their stockholders' money further, are actually considering preparing political risk assessments on investing in the US, as is done for so-called 'politically less stable' countries.

Global climate change

Global climate change – or 'global warming' as some characterize it – is a particularly challenging matter. Climate change presents the possibility of sweeping legislation and regulations, such as carbon taxes, that would probably have little positive effect on the environment. In 1992, partially in response to the perception that the world's climate is growing warmer due to mankind's use of fossil fuels, the United Nations Conference on Environment and Development met in Rio de Janeiro. As a result, 160 nations signed several agreements, including the Framework Convention on Climate Change. This committed developed countries, such as the US and other members of the Organization for Economic Cooperation and Development, to take action to return their emissions of greenhouse gases to 1990 levels. Developing countries, which will produce between 80 and 85 per cent of the total growth in greenhouse gas emissions over the next century, were initially required only to take an inventory of their greenhouse gas emissions.

One may question the need for precipitate action, since it cannot be determined how big a problem global warming is, or whether it even exists. The scientific community remains divided over whether the world's climate is changing, or simply experiencing normal cyclical fluctuations; or, if it is changing, whether it is warming or cooling; and if it is changing, how much human activity is contributing to this change.

What is known is that more than 90 per cent of the greenhouse effect comes from naturally caused water vapour. Only about 5 per cent of the greenhouse effect comes from carbon dioxide, and only between 2 and 3 per cent of that is man-made, from automobiles, smokestacks, etc. Put another way, nature accounts for 200 billion tons of carbon (as carbon dioxide) exchanged between the surface and the atmosphere each year; human activities release only 7 billion tons.

Nevertheless, a carbon tax, which many environmentalists are proposing, would be aimed at reducing that relatively tiny 7 billion tons, and would ignore the vast majority of greenhouse gases, which are produced by nature. The cost? A carbon tax of $100 per ton, according to US government studies, could cost the average American family $1,000 a year, drag the US economy down by 1-3 per cent of GNP by the year 2000, and cost 600,000 jobs. The benefit? There is considerable debate here as well.[5]

The US government recognized these realities when, in response to the Rio conference, it adopted a policy that entails taking sensible and realistic actions that slow down the rate of greenhouse gas emissions, but that stand on their own economic merits and do not lead to social and economic dislocation. These include conserving energy and improving energy efficiency, particularly the efficient use of fossil fuels, and are referred to as 'no regrets' policies.

---

[5] 'Limiting Net Greenhouse Gas Emissions in the United States', US Department of Energy, Washington DC, September 1990.

Politics, economics and environment:

The challenge

The financial resources of any nation – no matter how prosperous – that are available to undertake programmes are clearly finite. The US budget deficit, the costs of German reunification, and the slowdown of the Japanese economic boom all demonstrate that point. As the US experience has shown, environmental regulations can get out of control, diverting resources that could be needed to address other, more pressing social needs. Global leaders of today and tomorrow must ask: how do we control the direction and usefulness of future environmental, health and safety requirements, so they do not waste those precious resources? How do we balance the need for environmental protection with the need for economic growth?

There have been some encouraging signs recently of a growing understanding of this in Washington. In August 1993, a bill was introduced in Congress, 'The Risk Communication Act of 1993'. It would require federal agencies to provide fair, scientifically sound and consistent assessments of potential health, safety or ecological risks before establishing regulations. And in October 1993, President Clinton gave federal agencies greater authority to write regulations on pollution and various business practices, and *required* them to analyse the costs and benefits of such regulations. Perhaps some 'dirt-eating' regulatory abuses can be avoided.

Summary of principles

There is obviously no simple answer to complex issues of environmental protection and enhancement. But policymakers must recognize that no country can afford to do everything it would like. Instead, they must set priorities on which problems are addressed in what order. Texaco offers the following guiding principles, which we believe will be useful to policy makers in formulating appropriate legislation and regulation to address environmental and energy issues:

- Environmental policies should be based, as far as possible, on scientific evidence, not on political pressure. Continuing scientific research and analysis to reduce uncertainties is critical, particularly in the field of climate change.

- Policy options should be subjected to rigorous cost-benefit analysis, and weighed carefully against all possible policy alternatives. Is the benefit society expects to gain from, for example, a carbon tax, worth the additional cost to millions of consumers? Or should other environmental concerns be targeted, where both the science and the demonstrated need for immediate attention is clear? Examples would be improving water and air quality, or cleaning up toxic waste sites.

- There are a number of practical steps, 'no regrets' options, that policy-makers can take now to protect the environment at a relatively low cost. These include improving energy efficiency and investing in conservation; both options at the low end of the cost curve. These measures to benefit the environment make sense under any conditions.

- On a global scale, free and open international trade encourages energy efficiency and conservation, and promotes environmental protection, as does a business climate that welcomes foreign investment, encourages technology development and cooperation, and protects intellectual property rights.

- On the negative side, ill-conceived and costly government mandates discourage both economic growth and environmental protection. Command and control measures distort economic growth and development, diverting resources from what consumers may prefer and what efficiency dictates.

The role of business

Protection and improvement of the environment is everyone's business. Industry must become a more active partner in the process, by ensuring that environmental considerations become an integral part of its day-to-day management and decision-making process. But business cannot afford simply to react to government initiatives. The oil industry has learned that every aspect of business must include environmental planning and development: exploration, research, production, storage, transportation, marketing, product use and disposal. As successful entrepreneurs in a competitive world, business leaders are eminently well qualified to help find and implement the solutions that society is looking for in protecting the environment. Indeed, business must participate in the political process of policy determination as well. In fact, it has both a right and an obligation to do so. Industry, unjustly seen by many as 'the problem', is in reality 'the solution'. It is, and will be, industry's capital, technology and management that will enhance and protect the global environment. Environmental policies should recognize this fact.

Sustainable Coal Use: the Role of Technology

Anthony Baker and John C. Whitehead

Abstract

Authoritative projections indicate that coal use over the next 30 years will continue to increase, particularly to supply the energy needs of developing countries. There will, therefore, be a continuing challenge to make coal use fully compatible with more stringent environmental requirements.

Already mining schemes, undertaken both in developed countries and by international mining companies generally, are designed to minimize the impact on land use, water and wildlife. Pollution from coal use (especially from the growing sector of power generation) has been reduced considerably over the last 25 years in developed countries throughout the world, by use of modern clean-up technologies. Advanced 'clean coal technology' in its various forms can reduce pollutant amounts to near zero, but the thrust to demonstrate and install such technologies widely is impeded by the fragmentation of interests in the coal-use chain.

A further advantage of newer coal use technology is the higher conversion efficiency to electrical energy compared with existing technology – and substantially higher than plant currently used in developing countries. An important part of any precautionary response to meet concerns about potential climate change from carbon dioxide emissions must lie in promoting the transfer of modern efficient coal use technology to developing countries.

Introduction

One difficulty of any discussion of 'sustainability' is the choice of an appropriate timescale. Some may wish to dwell on the period from the year 2050 onwards, and on that prospect *all* fossil fuels, even coal with proved recoverable reserves of over 200 years' supply on current usage, may be seen as having a transient role. However, even for those with crystal balls, such distant considerations beg many questions – about the nature of development of society in interdependence or conflict, and about the development of radically different technologies, not least in energy provision.

However, by adopting a shorter timescale, 20 to 30 years, a great deal can be effected to promote greater cohesion of economic activity with environment demands, and with the goal of inter-generational sustainability firmly in mind. That is our perspective in this paper. In this timeframe we acknowledge that much will be done to develop non-fossil energy forms, of types appropriate in scale of operation and cost, and desirably without their own environmental dangers. But over the next generation the effective exploitation of fossil fuels, coal in particular, will be crucial for the achievement of adequate economic well-being, a well-being without which any thoughts of longer-term transitions in energy provision may be illusory.

Coal use will grow

Our first proposition is, therefore, a backdrop to the whole of our paper. On a *world view*, coal use is growing and, on any reasonable assessment, will continue to grow over the next 20 to 30 years. As one illustration, Figure 1 shows the projections contained in the International Energy Agency (IEA) *World Energy Outlook* for summer 1993 – a growth projection of 50 per cent or so over the 20 years from 1990.[1] The World Energy Conference scenarios published in September 1993 are similar, to varying degrees depending on the scenario.[2] Any such projection will be wrong in detail, but two things seem virtually certain:

[1] IEA/OECD, *World Energy Outlook*, IEA/OECD, Paris, 1993.
[2] WEC Commission, *Energy for Tomorrow's World*, Kogan Page, London, and St Martin's Press, New York, 1993.

Figure 1 World coal demand (including lignite) 1971–2010, billion tonnes of coal equivalent

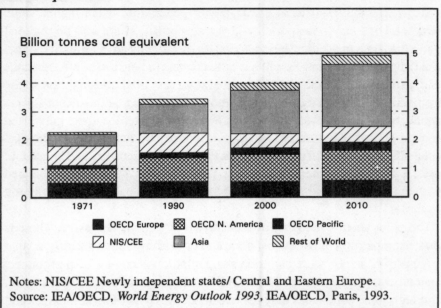

Notes: NIS/CEE Newly independent states/ Central and Eastern Europe.
Source: IEA/OECD, *World Energy Outlook 1993*, IEA/OECD, Paris, 1993.

- OECD Europe's coal use will be an even smaller fraction of total coal demand than now.

- The real growth will occur in the developing world: Asia, China and India in particular.

So our European policies for matching energy use to environment needs will be effective on the wider scale only in so far as they will have impact elsewhere, especially in Asia. That means that we in Europe cannot just dismiss coal from the energy balance.

The reasons for continuing growth in coal are well known: the need for energy to meet the needs and aspirations of increased population, economic growth and urbanization. The IEA in the same study, for example, predicts that world power demand will grow to around 21,000 terawatt-hours (TWh) by 2010 and that the proportion of coal-fired capacity will be maintained.

For relatively poor countries in particular, coal has much to recommend it. It is abundant, and relatively cheap to access and use. It is a low value energy source, well suited to local use in power generation, when higher value hydrocarbons may be exported with greater profit. There will be regional variations both in the increased demand for electricity and in terms of the fuel used; developing countries will have the greatest increase in power demands and also have access to significant coal reserves. Coal consumption in the iron and steel industry can be expected to move towards the use of lower-quality coals blended with coking coal. Further, the increased use of pulverized coal injection in blast furnaces will slightly reduce the demand for metallurgical coke. The use of coal in the heating market, including industry and household uses, is decreasing generally; however, it is likely that the demand associated with this type of use will continue to be significant, especially in developing countries.

The point about coal being of relatively low value is pervasive. There is fierce competition in world coal trade and prices have fallen steadily, so that large profits do not accrue to miners and mine-owners. Few coal producers operate with high margins. The world's coal industries are in general poor, although some companies may operate under the cover of oil companies and of such conglomerates as the Hanson Corporation.

Environmental challenges

With coal firmly established on the world scene, there will inevitably continue to be environmental challenges to its production and use, particularly as environmental legislation tightens to reflect society's aspirations. The Box overleaf lists the principal legislation which impacts on coal supply and use in the UK, not a country especially burdened by legislation by comparison with, say, the US. There are distinct kinds of challenge posed by coal use, and it may be helpful to discuss them under three headings, as below.

Local challenges to minimize mining impact
In all developed countries now there are major challenges to coal developers and mining companies in opening up new mines and operating existing ones. We know this only too well in the UK: British Coal failed to secure permission

to mine coal in Warwickshire in recent years; its Leicestershire mining plans were severely curtailed; it is experiencing increasing difficulty in getting agreement to work new opencast sites in England.

However, new mines continue to be worked in other developed countries, notably the US and Australia, and by international mining companies in low-cost areas in developing countries, but only after full environmental impact statements have been made and under strict conditions and guidelines.

It is impossible to estimate a standard cost of such compliance for two reasons. First, circumstances are highly local. Contrast the new mines at Selby in Yorkshire ten years ago, incorporating water tables, consideration of visual impact, the need to avoid spoil heaps, with new mines in Indonesia which are characterised by 100-metre seams of opencast coal and low overburden, but which have much greater infrastructure need. Secondly, a high technology mining scheme will be planned to meet environmental guidelines from the start; for example:

• To minimize spoil extraction
• To provide adequate water treatment
• To provide site screening and subsequent reclamation.

New mine sites are designed to be rehabilitated to a state equivalent to and sometimes more attractive than that prior to development. Surface and ground water are protected by secure handling and disposal of toxic and acid materials, and mining and reclamation procedures are designed to minimize water pollution; water quality is constantly monitored. Solid and rock extracted in gaining access to the coal are disposed of in structures engineered to ensure stability, and topsoil removed from mining areas is stockpiled and subsequently replaced. Land uses after mining has finished include pasture, agriculture, and the creation of wildlife habitats.

The problem comes in getting older mines to comply with latest planning and environmental guidelines – and it is obvious that there have been and will continue to be tensions. But the message is clear for mine operators: comply pragmatically, or cease operating.

Existing and proposed legislation in the UK

Existing legislation

- *EC Large Combustion Plant Directive (LCPD)* Regulates emissions of SO_2, NO_x and particulates from boilers and furnaces larger than 50 thermal megawatts (MWth). To be revised in 1994.

- *UK Environmental Protection Act (EPA)* (1990) Framework legislation concerned with wide range of environmental issues, resulted in specific legislation for industrial processes. Examples include processes authorized centrally by Her Majesty's Inspectorate of Pollution with regard to integrated pollution control.

  There are also similar guidance notes for local authority control of air pollution for smaller, less complex processes and including legislation for boilers with capacities 20–50MWth.

- *UK Clean Air Act* Incorporated into the EPA but regulates small coal-fired plant (less than 20MWth) with regard to emissions of smoke and grits.

- *Air Quality Standards for SO_2, NO_x and smoke* Originated as EC Directives but now incorporated into UK legislation.

- *Control of Pollution Act and Special Waste Regulations* Now incorporated into the EPA. Contains regulations restricting the disposal of certain categories of waste (including combustion of solid residues).

- *Special Waste Regulations* Currently mining waste is exempt as a special waste, but the UK Department of Environment is consulting on redefinition of certain categories of waste. Combustion residues containing lime (FGD, FBC with limestone addition) classified as special waste and require specific licence.

- *The Framework Convention on Climate Change* Signed at Rio in 1992. Commits the UK to reduce emissions of greenhouse gases to 1990 levels by the year 2000. In January 1994 the government issued its strategy document for reducing emissions of CO_2. No developments on other greenhouse gases as yet.

- *EC proposals for a carbon/energy tax.*

- *Eco labelling* Industrialists who burn coal as part of their process may be prejudiced due to the negative score associated with emissions of SO_2.

- *United Nations Economic Commission for Europe (UNECE) Protocols on SO_2 and NO_x* UK government is a signatory to these protocols which aim to reduce emissions of SO_2 and NO_x. New protocols were effectively agreed in March 1994.

- *Critical loads* A method of identifying areas at risk from acid deposition. The critical loads concept is to be applied in redefining UNECE targets for SO_2 and NO_x, and is used in the UK in determining annual emission budgets of these gases for industrial sites.

- *Control of Substances Hazardous to Health (COSHH)* Has implications in all aspects of production, utilization and disposal of coal.

- *Duty of care* Places the responsibility on the individual to ensure safe disposal of waste.

Proposed legislation

- *Small Combustion Plant Directive* Institute Français de l'Energie study recommended emissions legislation for combustion plant down to 1MWth.

- *EC Landfill Directive* To determine who is qualified to dispose of waste, and the type of waste that can be disposed to landfill.

Pollutant emissions: paying for the current technologies

The largest regional challenge to coal use is in controlling pollutant emissions from coal burning: particulate material, NO_x and SO_2 in particular. While around 15 per cent of OECD coal use is in iron and steel making, the growth sector is power generation. In that area, there is continuing tension between the tightening of environmental standards in all developed countries (representing the goal of the common good) and the costs which coal-using electricity utilities have to pay in order to comply.

And their costs are considerable. In the UK National Power and PowerGen have referred, in their recent environmental reports, to their combined investment of £1.4 billion at existing plants to provide flue gas desulphurization (FGD) at some power stations and low-NO_x burners more generally.[3] As a round figure, FGD might be an add-on cost of around 0.5p/kWh to any *existing* power plant, but much lower if designed in from the beginning. Extra tension derives from the fact that the investment has to be made by the power generator, who will, not surprisingly therefore, seek lower coal prices from the coal producer, who is already likely to be operating at low margins.

Despite such tensions, a great deal has been done over the past 15 years to make coal-fired power stations a great deal less polluting. Most has been done *outside the UK* where for example, the US, Japan, Germany and Denmark all operate modern coal-fired plant to 'new source' emission standards. There it has been either directly in the interest of the utilities, or of the utilities and coal industries in close partnership, to make the necessary investments for the medium term, rather than fight environmental concerns and the requisite changes. There have also been significant developments in technologies which are now available and employed to reduce dramatically emissions of SO_2 and NO_x. In the US, for instance, there has been a 25 per cent reduction in SO_2 emissions since 1970 (when the first Clean Air Act was introduced), over which period there has been a 60 per cent increase in coal use. This process goes on – especially where new plant is being commissioned – in Japan or Germany, for example.

---

[3] National Power plc, *National Power Environmental Performance Report 1992*, National Power plc, London, July 1993. PowerGen plc, *Environmental Performance Report – Analysis of PowerGen's Achievements*, PowerGen plc, London, July 1993.

Clean coal technologies: the challenge of cohesive development

The further challenge is how to bring about an investment in new breeds of coal power plants, that can substantially reduce NO_x and virtually eliminate SO_2. Here there is a variety of approaches developing around the generic forms of pressurized fluidized bed combustion or coal gasification, each integrated with combined cycle power generation, using gas and steam turbine technology. An IEA Coal Research report gives more details of such technologies.[4]

Figure 2 summarizes how the newer processes of pressurized fluidized bed combustion (PFBC) (with added limestone to absorb sulphur) and integrated gasification combined cycle plant can dramatically reduce the pollutant burden, even compared with the modern pulverized fuel and FGD plant, itself considered a 'clean coal' plant in most developed countries. The extra advantage of the newer types of plant is their anticipated higher conversion efficiencies. Perhaps a further 8 percentage points, over and above the roughly 40 per cent conversion rates achievable with the best conventional plant now, is available from the combined cycle operation. In this way the extra environmental control is available at no cost to efficiency. This is a point clearly in favour of these advanced processes compared to FGD installation and other clean-up technology which are often criticized for reducing the efficiency of conversion.

Such higher efficiency plant *should* produce electricity more cheaply than conventional plant with added controls, but that has yet to be *demonstrated*. Putting the necessary programmes in place is a major challenge. The US clean coal programme has, for example, a $5 billion bill attached to it, and it is clear that such programmes require firm cohesive support from governments, electric utilities and coal industries, working to a common end. The work of a body such as the US Electric Power Research Institute is of enormous value in focusing the thrust of such efforts.

The great difficulty, perhaps outside the US and Japan and Germany, and with honourable exceptions such as the Buggenum IGCC (integrated gasification combined cycle) plant in the Netherlands or the ABB work in

---

[4] Chris Maude, *Advanced power generation – a comparative study of design options for coal*, Report IEACR/55, IEA Coal Research, London, March 1993.

Figure 2 Coal-fired plant emissions, kg/MWh

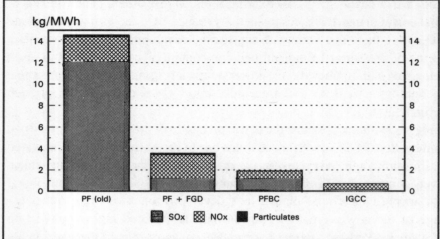

Notes: PF=pulverized fuel; FGD=flue gas desulphurization; PFBC=pressurized fluidized bed combustion; IGCC=integrated gasification combined cycle.
Source: Coal Research Establishment.

Sweden, is the fragmentation of interests which has so often impeded carrying the whole enterprise through. At the Grimethorpe plant in Yorkshire, for example, a great deal of information and experience has been gained about the operation of a pressurized fluidized bed plant, without as yet a thrust forward into a real power plant. Recently, however, an industry-led programme including GEC-Alsthom, British Coal, and PowerGen was announced to develop the Topping Cycle power generation system with funding from the UK Department of Trade and Industry and the European Union.

Coal suppliers have a clear interest in getting such plants built and the technologies adopted, but their influence is limited and they have no money, as coal prices are steadily forced down. Electric utilities have money, but in developed countries they often have more seductive technologies in which to invest. The shorter payback timeframes and implied higher discount rates of the world of privatized utilities are further deterrents.

So 'advanced clean coal technology' *is* being developed in Europe, in Japan and in the US – but it has been much slower in fulfilling the hopes many of us have held for the last 20 years.

Carbon dioxide: achieving effective precautionary policies

Some will see the real challenge as arranging the fastest and most complete demise of coal use in order to minimize the threat of possible global warming from CO_2. Other papers in this volume allude to the considerable doubts that remain, and are likely to remain for a long time, about the nature and extent of enhanced global warming and the role of CO_2, and we do not want to get into that debate here. Solutions involving CO_2 mitigation and capture technologies are now under investigation by, among others, the IEA Greenhouse Gas R&D Programme. Early indications are that the total capture and disposal/storage of CO_2 from a coal power plant would be expensive and might add 1.3–1.7p/kWh to the cost of power production.[5] That level of cost increase, while not impossible to contemplate, is unlikely to make the adoption of total CO_2 removal attractive.

Instead, it may be helpful to reflect again on the backdrop to this paper of the virtual certainty of great increases in coal use in developing countries. Even if coal use in Europe were to be sharply reduced, it would have little impact on CO_2 emissions from coal use generally. Here, in Europe, the real issue is how to limit CO_2 emissions in transport use; for coal the issue arises primarily in relation to developing countries in Asia. They do not share our concerns over global climate issues because they have greater imperatives for food and water and for economic development, with the inevitable energy use that goes with it at their stage of economic activity.

If we in Europe adopt policies which encourage electric utilities in the developed world to back out of coal use, one thing is certain. The adoption and spread of the clean, higher efficiency combined cycle technologies will be that much slower and more uncertain. Developing countries (where conversion efficiencies in electricity generation are perhaps 25 per cent at best) will produce considerably more CO_2 per unit of useful energy than they need do. *Instead, what we ought to be doing is to put in place the advanced clean coal technologies and in parallel develop schemes to spread them to developing countries.* If we seriously want to minimize CO_2 production in the next 20

---

[5] I. Webster, H. Audus, P. Riemer and W. Omerod, 'Carbon Dioxide Capture and Disposal as Greenhouse Gas Abatement Option', Paper to IEA Conference on Clean and Efficient Use of Coal and Lignite, Hong Kong, December 1993.

years, as opposed to promoting some alternative technology on other grounds, the spread of coal use with higher efficiency must be part of the policy.

Putting in place effective approaches to achieve such technology transfer is, of course, by no means easy, particularly as it includes balancing the private interests of technology developers with the wider good of lower CO_2 production. A large number of design studies for new plants are being undertaken, often funded under multilateral and bilateral agreements. These provide information on the requirements for new coal-based technology in developing countries. In addition to various national initiatives, both the European Union and the International Energy Agency have clearly stated intentions to assist developing countries with the provision of new technology. However, the practical difficulties in financing this intention are immense and will require continuing concerted efforts.

Moreover, some apparently rational policies may be counter-productive: for example, whatever may be the theoretical arguments for carbon taxes in the developed countries, such taxes will discourage the spread of more efficient coal use. A tax on *wasted energy* would be another matter.

In seeking to achieve more efficient and less polluting technologies for using coal, coal industries will play their part – but policymakers must ensure a framework which includes a place for advanced coal technology. To those who are serious about the immediate need to combat the risks of global warming, we strongly propose that this is a key part of any effective precautionary policy. It is practical, achievable and has mutual benefit to developing countries and to ourselves in Europe.

Conclusions

- With large recoverable reserves, widespread availability and requiring only low investment for its exploitation, coal use on a world scale is set to grow over the next 30 years. This is particularly so in Asia. The challenge is to use coal cleanly and with maximum efficiency.

- Modern mining schemes being undertaken in developed countries and elsewhere by large international mining companies are implemented with strict environmental controls and safeguards. The continuing tensions relate to historic mining liabilities at older or former mines.

- In coping with the main potential pollutants of coal use – NO_x, SO_2 and particulate material – much has already be done by technology-based controls to meet the demands of ever-tightening legislation. Further improvements are available, especially where coal producers and electric utilities can take concerted action.

- More advanced technologies, also employing combined cycle power generation, offer the opportunity to minimize pollutant emissions, while at the same time increasing the efficiency of power generation. Putting such technologies in place is costly, and fragmentation of interests has made for relatively slow development.

- Given the inevitable worldwide growth of coal use in the next decades, any practical response to fears of enhanced global warming through CO_2 emissions must include policies to promote the adoption of such higher efficiency technologies, in developed countries at first, and then a rapid spread to developing countries.

The Pros and Cons of Learning by Doing: the California Utilities' Experience in Energy Efficiency and Renewables

Fereidoon P. Sioshansi

Abstract[1]

Following the 1973 Arab oil embargo, California utilities, with the strong support and encouragement of their regulators, went on an unprecedented energy conservation and fuel diversification rampage. During the ensuing decade, millions of dollars of ratepayer-funded money were spent on conservation and load management (CLM) and a wide variety of renewable energy technologies. With the passage of the federal Public Utilities Regulatory Policy Act of 1978 (PURPA), a new generation of non-utility generators (NUGs) and independent power producers (IPPs) entered the market, bringing an element of competition previously unknown to the industry.

The lessons learned from these 20 years of experimentation are instructive to anyone interested in legislating on energy policy. The experience clearly shows what can – and cannot – be done efficiently through regulatory mandates. It also illustrates how well-intentioned regulations may sometimes produce unexpected and counterproductive results. Overall, the California experience is a mixed one. Some mid-course corrections were necessary when it became clear that the end would not justify the means.

This paper focuses on two particular aspects of the experiment: energy conservation and renewable energy. In each case, it will review what was originally intended, what actually happened, and how things evolved over

---

[1] The views expressed in this paper are those of the author and do not necessarily reflect the official position of EPRI or its member utilities.

time to their current position. Useful lessons are drawn from this experience when appropriate.

In the case of energy conservation, it is clear that if you remove the disincentives and provide the utilities with sufficient financial incentives, they will pursue energy efficiency options. But there is an ongoing debate as to whether ratepayer-funded demand-side management (DSM) is the best way to finance energy efficiency and whether utilities are the best vehicles for delivery of such services.

As for renewable energy, California regulators, through their aggressive promotion policies, placed a de facto tax on electric utility ratepayers in the state to fund a substantial amount of research, development and demonstration (RD&D) spending and experimentation which are now bearing fruit in California and elsewhere. The relevant policy question is not whether they achieved the intended results, but whether it was an efficient and equitable way to finance the RD&D effort.

Introduction

As countries around the world are becoming increasingly interested in clean, efficient use of energy, and renewable forms of energy, a fundamental question on the minds of many policymakers is 'What is the best way to promote efficient use of energy by consumers and energy suppliers and to promote the introduction of renewable energy technologies?'. In countries with high energy prices and taxes, the market forces provide sufficient economic incentives for energy conservation and some policymakers believe that the role of the government should be limited to removing major barriers to the market mechanism and/or educating the consumers in the benefits of conservation. Some studies suggest that, while there are some market 'imperfections', there are no insurmountable market 'failures' in energy conservation.[2] Furthermore, even when certain market imperfections exist, it is not clear whether government-mandated demand-side management (DSM) measures are necessary or effective.[3]

Other studies argue that there are major distortions in the energy markets which explicitly or implicitly favour conventional energy generation and use, and disfavour investment in renewable energy or efficient energy appliances.[4] Those subscribing to this theory suggest that governments have not only a *right* but an *obligation* to intervene in the energy market to redress this imbalance. The proponents of this theory fall into two broad, and sometimes overlapping, camps: one group espouses the 'stick' method, i.e. force the market to behave in certain ways using the visible fist of regulations; while the other group espouses the 'carrot' method, i.e. provide financial rewards (or penalties) to energy producers and/or consumers to modify their energy consumption behaviour using the invisible hand of economic theory.[5]

---

[2] Ronald J. Sutherland, 'Market Barriers to Energy Efficiency Investments', *Energy Journal*, Vol. 12, No. 3, 1991, pp. 15–34.

[3] Albert Nichols, 'How Well Do Market Failures Support the Need for DSM?', National Economic Research Associates, Cambridge, MA, August 1992.

[4] Anthony Fisher and Michael Rothkopf, 'Market Failure and Energy Policy', *Energy Policy*, Vol. 17, No. 4, 1989, pp. 397–406. T. Scott Newlon and David Weitzel, 'The Evidence for Market Failure as a Motivation for Utility-sponsored Conservation Programs', National Economic Research Associates, Washington DC, April 1990.

[5] R. Cavanaugh, 'Responsible Power Marketing in an Increasingly Competitive Era', *Yale Journal of Regulation*, Vol. 5, No. 331, 1988, pp. 331–358.

Each of these three approaches has its proponents and critics, and there is little empirical evidence to discredit any one approach. So the debate on which approach produces the best results with least cost and economic distortion continues. As it turns out, several utilities in California have been experimenting with elements of the three theories and some conclusions may be drawn on the relative merits of each, even though we are far from a unanimous consensus on what works, or works best. The interest in what may be learned from their experience is more than academic. This paper is an attempt to review the California experience in promoting energy conservation and renewable energy and to draw some tentative conclusions on what can be learned from the still evolving experiment.

The paper is organized into two parts. The first deals with energy conservation, sometimes labelled demand-side management (DSM) in the US literature. The second is focused on attempts to promote renewable energy.

PART ONE: ENERGY CONSERVATION

Restraining demand: vintage 1970

Prior to the 1970s, utilities in the United States were left pretty much alone to decide how (and whether) they handled their marketing. Only a few states specifically prohibited certain types of marketing in their jurisdiction.

This *laissez-faire* situation came to an abrupt end following the 1973 Arab oil embargo, when federal and state regulators became concerned about the rapidly rising costs and the availability of energy supplies. Customers in regions such as California which were heavily dependent on imported oil for power generation were shocked to see their electricity rates rise following years of relatively stable prices. There was a rude awakening for some utilities such as Southern California Edison Company (SCE) which was dependent on oil- and gas-fired generation for 67 per cent of its energy in 1973 when the price of oil quadrupled.[6]

---

[6] Fereidoon P. Sioshansi and S.J. Nola, 'Role of the US Utility Industry in the Commercialization of Renewable Energy Technologies for Power Generation', *Annual Review of Energy*, Annual Reviews Inc., Palo Alto, CA, Vol. 15, October 1990, pp. 99–119.

One result of the embargo was a flurry of federal and state policies including President Carter's National Energy Plan which contained the now famous Public Utility Regulatory Policies Act (PURPA) of 1978. The Act's objectives were to diversify the nation's energy supplies to include an increased reliance on domestic renewable energy sources, to encourage competition in the electric power sector, and to promote energy conservation.[7] A number of initiatives to set appliance efficiency standards and strengthen building codes also followed.

At the state level, California went even further by establishing its own energy policies. California's experience is particularly noteworthy not only because of its sheer size, but also because of its status as a national trend-setter.

The two primary regulatory agencies in California, the California Energy Commission (CEC) and the influential California Public Utilities Commission (CPUC) set an aggressive conservation agenda for the utilities in the state starting in the mid 1970s. The CPUC, for example, established spending targets for conservation and load management (CLM) for the state's four large investor-owned electric and gas utilities.[8] The companies were, in effect, required to spend a mandated quota on CLM. Not surprisingly, the CLM spending of the utilities in the state took off. (See Figure 1.). During this early period, neither the regulators nor the utilities knew for sure what they could accomplish, nor did they know the full impact of the CLM programmes on their customers' rates. It was a classic case of learning by doing. As it turned out, the initial results were quite impressive.

Two things contributed to the success of the early CLM programmes in California in the 1970s and early 1980s as measured by their reported cost-effectiveness. First, the public was sensitized to the need for energy conservation and willing to participate in utility-sponsored programmes. It was unpatriotic not to. Second, since generations of Americans had been raised on abundant and cheap energy (e.g. a gallon of gasoline cost 30¢ before

[7] Fereidoon P. Sioshansi, 'Promise vs.Reality of "Soft" Energy', *FORUM for Applied Research and Public Policy*, Vol. 7, No. 2, University of Tennessee, Knoxville, TN, Summer 1992.
[8] California's four investor-owned utilities are Pacific Gas and Electric (PG&E), Southern California Edison Company (SCE), Southern California Gas Company (SoCal), and San Diego Gas and Electric (SDG&E).

Figure 1 California utilities' conservation funding trend 1980–92, $million

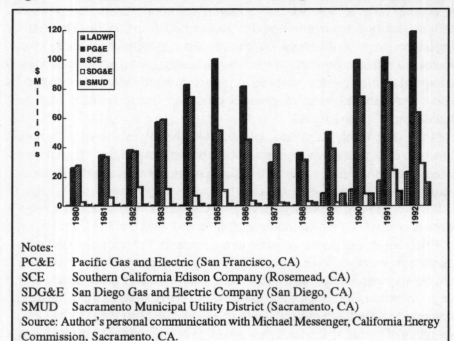

Notes:
PC&E   Pacific Gas and Electric (San Francisco, CA)
SCE   Southern California Edison Company (Rosemead, CA)
SDG&E   San Diego Gas and Electric Company (San Diego, CA)
SMUD   Sacramento Municipal Utility District (Sacramento, CA)
Source: Author's personal communication with Michael Messenger, California Energy
Commission, Sacramento, CA.

1973), the opportunities for painless conservation (e.g. turning off the lights
in unoccupied rooms!) were enormous and could be had at little or no cost.
The economic principle of diminishing marginal returns did not apply until
much later. Subsequent energy price increases, leading to the all-time oil price
peak of $40 per barrel in 1980 reinforced the *need* for, and the *cost-
effectiveness* of, energy conservation.

But the unexpected downturn in energy prices, beginning in 1981, reduced
the urgency and the cost-effectiveness of CLM programmes. Furthermore,
during the Reagan–Bush Administration, both the federal and state bureaucrats
lost their zeal in energy conservation. Energy-related research budgets were
slashed as the Department of Energy (DOE) scaled back its R&D programmes

and gave the market forces free rein.[9] The combined effect of these developments may be seen in Figure 1. By 1988, the California CLM spending was scaled back to half of its 1984 peak. More depressing was that the cost-effectiveness of the programmes had fallen to one-quarter of their early 1980 peaks as measured by MWh saved per dollar spent.[10] Something had gone wrong. It was time for a different approach to promote energy conservation.

The dawn of the negawatt era

The fundamental question, posed by a CEC insider, was 'Will electric utilities effectively compete in markets without a profit motive?'. The answer was fairly obvious by the late 1980s: not effectively, not aggressively, nor willingly![11] The reasons were equally obvious. Profit-motivated investor-owned utilities lose on several grounds when they engage in CLM:

- Selling fewer kilowatt-hours means that the utility's substantial fixed costs have to be divided among fewer kWhs, hence increasing the rates.

- Energy conservation programmes cost money to implement, further adding to a utility's costs, but not necessarily to revenues.

- For utilities with high retail rates and low marginal costs, CLM results in substantial 'lost revenues'.[12]

- For utilities with excess capacity and/or energy, CLM does not make near-term economic sense.

- Without financial incentives to utilities, there is no way to make money for the utility shareholders by pursuing demand-side options even if the expenses are covered.

---

[9] Roland J. Sutherland, 'An Analysis of the US Department of Energy's Civilian R&D Budget', *Energy Journal*, Vol. 10, No. 2, 1989, pp. 35–54.
[10] Michael Messenger, 'Will Electric Utilities Effectively Compete in Markets Without a Profit Motive?', California Energy Commission, San Francisco, CA, May 1989.
[11] Ibid.
[12] In California, for example, where retail rates are around 10¢/kWh and marginal costs in the range of 3¢/kWh, the utility would 'lose' 7¢/kWh for every kWh not sold (also known as a 'negawatthour').

So, the next question, also posed by the same CEC insider, was 'Can greed, accuracy, and fairness be mixed for the public good?'.[13] The answer to this became a de facto *yes* in the late 1980s and has been endorsed in varying degrees by a growing number of state regulators across the United States.[14]

DSM defined

Over the years, demand-side management has assumed many connotations. The following from *Utilities Policy* is typical:[15]

> DSM, broadly defined, includes any initiative undertaken by the utility – with the customer's cooperation and consent – that provides the customer with essentially the same level of energy services but at lower overall costs. Many, but not all, DSM schemes involve some form of economic incentive in exchange for a different energy usage pattern. This definition includes such things as energy conservation (i.e. getting the same level of energy services with fewer kilowatt hours), peak clipping, valley filling, strategic load building (e.g. through interruptible rates or other pricing incentives), and selected fuel substitution, so long as the customer is made better off and there are no net environmental or social adverse effects.

Economics of DSM

There are three basic reasons why profit-motivated utilities shy away from DSM in the absence of financial incentives:

• They (generally) lose revenues when they sell fewer kWhs

---

[13] Michael Messenger, 'Can Greed, Accuracy, and Fairness be Mixed for the Public Good?', California Energy Commission, San Francisco, CA, August 1992.

[14] At the last count, 15 states and the District of Columbia have DSM incentive mechanisms in place; another 9 are considering such incentives; and 5 have approved generic incentives in principle, but have not specified exactly how they would be implemented. For a discussion of DSM incentives, see 'Shaping DSM as a Resource', *EPRI Journal*, October/November 1991, Palo Alto, CA.

[15] Fereidoon P. Sioshansi, 'Demand-Side Management and Environmental Externalities: Ramifications on Utility Resource Planning', *Utilities Policy*, October 1992, pp. 320–329.

- They incur out-of-pocket expenses to plan, administer and implement DSM programmes; but cannot (generally) recoup the full extent of their expenses from the short-term savings resulting from DSM.

- There is some risk but (generally) no reward in DSM for the shareholders.

The first two are obvious disincentives that had to be removed; the third obstacle requires a reward. To make utilities enthusiastic about DSM, regulators must first remove the disincentives, and then add some incentives to sweeten the deal. Removing the disincentives is necessary but not sufficient. These simple observations were not immediately obvious to the regulators, although they were to the utilities. Fortunately, several key intervenor groups came to the rescue in exercises that have been code-named 'collaborative processes' in the northeast US, California, and elsewhere.[16] The consensus that has emerged out of these exercises has basically acknowledged the existence of the three obstacles to DSM listed above and the need to remove them.[17] At the last count, 24 states and the District of Columbia (see Table 1) have partially or completely removed the obstacles and/or provided some form of incentive to utilities in their jurisdictions for undertaking aggressive DSM programmes.[18]

Three parallel developments have further enhanced the *righteousness* of DSM. First, many states now require utilities to do integrated resource planning (IRP) or least cost planning (LCP), which is another way of saying that they want the utility to show a good-faith effort to consider demand-side options alongside supply-side options in its resource plan.[19] Second, the growing concerns about environmental degradation caused by generation and transmission of electricity have forced many utilities to reconsider their future

---

[16] J. Raab and M. Schweitzer, *'Public Involvement in Integrated Resource Planning: A Study of Demand-Side Management Collaboratives'*, ORNL/CON-344, Oak Ridge National Laboratory, Oak Ridge, TN, 1992.
[17] David Moskovitz et al., 'Weighing Decoupling vs. Lost Revenues: Regulatory Considerations', *Electricity Journal*, November 1992, pp. 58–63. S. Wiel, 'Making electric efficiency profitable', *Public Utilities Fortnightly*, Vol. 124, No. 1, 6 July 1989, pp. 9–16.
[18] Edison Electric Institute, 'State Regulatory Developments in Integrated Resource Planning', Edison Electric Institute, Washington DC, 1992.
[19] Ibid.; Cynthia Mitchell, 'Integrated Resource Planning Survey: Where the States Stand', *Electricity Journal*, May 1992, pp. 10–15.

Table 1 US DSM shareholder incentives

State	Shared savings	ROE bonus for DSM	Adjustment to overall return	Bounty to shareholders per unit	Mark-up on expenditures
Arizona	o	N	N	N	N
California	o	N	N	N	o
Colorado	o	N	N	N	N
Conneticut	N	o	N	N	o
DC	o	N	N	N	N
Hawaii	o	o	o	N	o
Idaho	N	N	o	N	N
Indiana	o	N	N	N	N
Iowa	o	N	N	N	N
Kansas	N	o	N	N	N
Maine	o	N	N	N	N
Maryland	o	N	N	N	N
Massachusetts	N	N	N	o	N
Michigan	N	N	o	o	N
Minnesota	o	N	N	o	N
Monatana	N	o	N	N	N
New Hampshire	o	N	N	N	N
New Jersey	o	N	N	o	N
New York	o	N	o	N	N
Ohio	o	N	N	N	N
Oregon	o	N	N	N	N
Rhode Island	o	N	N	N	N
Texas	N	N	o	N	N
Vermont	o	N	N	N	N
Washington	N	N	o	o	N

Notes: o='applies'; N='no'; *Shared savings*: Allowing the utility to retain for its shareholders a predetermined portion of any savings realized through the use of DSM; *ROE adjustment*: Adjusting non-taxable allowed returns on equity (ROE) to reward (or penalize) utilities for relative progress in developing DSM potentials; *Rate base premium*: A premium for rate based DSM investments (i.e. a return premium over and above the return allowed on supply-side investments); *Performance premium*: A 'bounty' per unit (kW, kWh) of resource saved in excess of some nominal goal; *DSM mark-up*: A fixed mark-up (e.g. per cent or flat rate) on DSM expenditures.
Source: Edison Electric Institute, *IRP in the States: 1992 Sourcebook*, EEI Rate Regulation Dept, June 1992.

Table 2 Integrated resource planning methods: requirements for considering environmental extenalities

State	Qualitative consideration	Quantitative consideration		
		Explicit monetization	Adder/ discount	Rankings/ point
Arizona	x			
California		x	x	
Colorado				x
Connecticut	x			
Hawaii	x			
Illinois	x			
Iowa	x		x	
Massachusetts		x		
Nevada		x		
New Jersey		x	x	
New York			x	
Ohio	x			
Pennsylvania	x			
Texas	x			
Vermont			x	
Wisconsin			x	

Notes: Minnesota joined the ranks of states with monetized externality requirements in March 1994.
Source: Edison Electric Institute, *IRP in the States: 1992 Sourcebook*, EEI Rate Regulation Dept, June 1992.

resource options. Some have decided that in today's environment it is simply unpopular even to think about traditional supply-side options. Finally, several states are now giving serious consideration to internalizing the externality costs of power generation (see Table 2), further improving the economics (and popularity) of DSM and renewable options.[20]

The combined effect of these economic, social, environmental, and regulatory pressures has turned DSM into a growth industry. According to a recent

[20] Fereidoon P. Sioshansi, 'Demand-Side Management and Environmental Externalities', op. cit.

industry survey published in *Electric Light & Power* (January 1993),[21] the US power industry invested $550 million in capital expenditures in DSM in 1992, with another $2 billion in non-capital expenditures.[22] The same survey projects the industry's 1993–7 capital investment in DSM to exceed $4.7 billion. Other estimates project the industry's annual DSM expenditure (most of which is not capitalized) to reach $5 billion by 1995, and perhaps double that amount by the year 2000. Table 3 (overleaf) shows some of the big current DSM spenders, based on their 1991 DSM spending.[23]

Why are they selling negawatts?

Overseas observers sometimes consider the scale and scope of the US utility-sponsored DSM programmes counter intuitive. They want to know:

- Why are the US utilities selling negawatts?
- Doesn't it raise their rates?
- Who pays the higher electricity rates?
- Are customers happy with DSM?
- Are negawatts cost-effective?
- What will be the long-term impact of DSM on load growth?

The following is a synopsis.

Why are the US utilities selling negawatts?

DSM is strongly encouraged or mandated through integrated resource planning (IRP) requirements in many states, and is the politically right thing to do. Furthermore:

---

[21] 'Electric Utility Forecast: 1993 Capital Spending to Surge 15%', *Electric Light and Power*, January 1993.

[22] Since most utility DSM expenditures are 'expensed' (as opposed to 'capitalized'), the $552 million *capital* investment may be viewed as the tip of the iceberg ('expensed' meaning that the costs were fully recovered – i.e. collected in rates – in the year in which they were incurred; for example fuel costs).

[23] For a good summary of the current status of US utility-sponsored DSM programmes, see Eric Hirst, 'Electric Utility DSM Programs: Cost and Effect: 1991 to 2001', Oak Ridge National Laboratory, ORNL/CON-364, Oak Ridge, TN, May 1993, and Steve Nadel, 'Utility Demand-Side Management Experience and Potential – A Critical Review', *Annual Review of Energy and Environment*, Palo Alto, CA, 1992.

- DSM is strongly endorsed by the environmental movement (while 'traditional' supply-side options are clearly out of favour at present).

- DSM programmes, *if* properly planned, implemented and maintained, *can* increase customer satisfaction/welfare, and *can* be cost-effective, particularly relative to new resource options.

- Utilities in some jurisdictions can actually make money by investing in DSM.

As will be explained below, the qualifiers – *if* properly planned, implemented, and maintained – are the key to the success or failure of DSM programmes.

Doesn't it raise their rates?

Arguments of conservation gurus notwithstanding, most DSM programmes cost money to plan, implement and maintain. Administrative, promotional, evaluation and reporting requirements further add to the costs.[24] Although it is not easy – and unfair – to generalize, *most* DSM programmes result in somewhat higher average rates. Customers who participate in programmes tend to see their bills go down (even though their average *rates* usually rise); the non-participants tend to see both their bills *and* rates go up.[25]

Who gains, who loses, and what's fair become convoluted points as utilities, through their DSM programmes, become *de facto* tax collection and distribution agencies. Opponents of DSM programmes see this as an evil – violating the all-important notion of consumer sovereignty – since non-participants do not, in general, have the option of *not* paying for DSM programmes undertaken by the utility that serves them.[26] Large industrial

---

[24] L.G. Berry, 'The Administrative Costs of Energy Conservation Programs', ORNL/CON-294, Oak Ridge National Laboratory, Oak Ridge, TN, 1990.

[25] C.J. Cicchetti and W. Hogan, 'Including Unbundled Demand–Side Options in Electric Utility Bidding Programs', *Public Utilities Fortnightly*, 8 June 1989, pp. 9–20. F. Sioshansi, 'The Myths and Facts of Energy Efficiency: Survey of Implementation Issues', *Energy Policy*, Vol. 19, No. 3, 1991, pp. 231–43.

[26] Larry E. Ruff, 'Least-Cost Planning and Demand-Side Management: Six Common Fallacies and One Simple Truth', *Public Utilities Fortnightly*, 28 April 1988. Larry E. Ruff, 'Equity vs. Efficiency: Getting DSM Pricing Right', *Electricity Journal*, November 1992, pp. 243–35. Andrew Rudin, 'Negawatts Will Never Be Too Cheap To Meter', *Proceeding of the 5th National DSM Conference*, 30 July – 1 August 1991, Boston, MA, pp. 79–87.

Table 3 The top US utilities with the greatest 1991 DSM expenditures

Utility	State	Generation/Revenue		1991 DSM Programmes			1991 DSM as a % of		
		(GWh)	($ million)	Demand (MW)	Energy (GWh)	Cost ($ million)	Peak	Sales	Rev
Pacific Gas & Electric	CA	80427	7378	700	620	150.4	4.9	0.8	2.0
Southern California Edison	CA	78643	7292	2358	585	107.4	14.1	0.7	1.5
Conneticut Light & Power	CT	26386	2276	260	811	81.6	6.1	3.1	3.6
Consolidated Edison	NY	39227	4910	161	266	76.6	1.7	0.7	1.6
Florida Power & Light	FL	74331	5159	1132	2625	72.0	8.0	3.5	1.4
Florida Power	FL	29149	1719	998	408	58.6	16.8	1.4	3.4
Niagra Mohawk Power	NY	39371	2883	67	345	55.3	1.1	0.9	1.9
Massachusetts Electric	MA	15985	1364	108	341	53.7	3.7	2.1	3.9
Carolina Power & Light	NC	42771	2686	1318	4418	52.9	15.6	10.3	2.0
Duke Power	NC	74167	3817	1089	4	48.1	7.7	0.0	1.3
Paget Sound Power & Light	WA	21451	957	69	154	42.1	2.6	0.7	4.4
Wisconsin Electric Power	WI	26498	1293	290	853	40.3	6.1	3.2	3.1
Sacramento Municipal Utility District	CA	8746	644	277	60	38	12.8	0.7	5.9
Boston Edison	MA	15517	1315	92	275	37	3.5	1.8	2.8
San Diego Gas & Electric	CA	15862	1356	81	150	36.5	2.8	0.9	2.7
Long Island Lighting	NY	17806	2198	209	309	27.9	5.4	1.7	1.3
Virginia Electric & Power	VA	62166	3688	228	147	27.2	1.8	0.2	0.7

Cont.

Utility	State	Generation/Revenue		1991 DSM Programmes			1991 DSM as a % of		
		(GWh)	($ million)	Demand (MW)	Energy (GWh)	Cost ($ million)	Peak	Sales	Rev
Potomac Electric Power	DC	26379	1552	331	174	26.8	5.7	0.7	1.7
New York State Electric & Gas	NY	19605	1368	34	79	24.6	1.5	0.4	1.8
Public Service Electric & Gas	NJ	41651	3500	89	137	24.0	1.0	0.3	0.7
Los Angeles Dept of Water and Power†	CA	23853	1771	136	12	22.0	2.7	0.0	1.2
Seattle City Light†	WA	9371	283	27	238	18.7	2.2	2.5	6.6
Narragansett Electric	RI	4865	458	37	118	18.6	1.5	2.4	4.1
Texas Utilities Electric	TX	87354	4892	392	55	18.1	2.3	0.1	0.4
Union Electric	MO	38305	2089	215	17	18.0	3.1	0.0	0.9
Totals and Averages		919886	66845	10698	13200	1177.0	5.4	1.6	2.4

Notes: † Municipal utility; all others are investor-owned utilities (IOUs).
Source: Eric Hirst, 'Electric Utility DSM Program Costs and Effects: 1991 to 2001', Oak Ridge National Lab, TN ORNL-CON-364, Oak Ridge, May 1993.

customers, for example, often complain that they have to subsidize their less efficient competitors who qualify for certain DSM programmes.[27]

Proponents of DSM, on the other hand, point out that the society as a whole benefits and that this should be the overriding principle for deciding the cost-effectiveness of DSM.[28] Furthermore, they argue that by not pursuing even more expensive supply-side options, everyone will eventually benefit from utility-sponsored DSM programmes. The equity and fairness of DSM programmes are thorny, controversial issues that require tradeoffs between social good and individual consumer sovereignty.[29]

Who pays the higher electricity rates?

The customers do! In practice, the utility reaches an agreement with the regulators on what constitutes prudent expenses in their DSM programmes, and all reasonable costs, including management and administrative costs, are ultimately covered through higher rates to customers. And if there are DSM incentives (i.e. profits for utility shareholders), these come out of the customers' pockets too.

Are customers happy with DSM?

Since customers who participate in DSM programmes are volunteers, it is probably fair to say that most find something attractive in the programmes, since otherwise they wouldn't take part. The real question, however, is whether the non-participants – whose rates *and* bills tend to go up without any direct benefits – are happy with utility-sponsored DSM programmes. We suspect that most non-participants are less than happy with the prospect of paying higher rates to subsidize programme participants who benefit directly from lower bills.

---

[27] ELCON, the Electricity Consumers Resource Council, representing large industrial customers, has taken a vocal stand against utility-sponsored DSM programmes because of the involuntary nature of their 'financing' through raising the utility's average rates.
[28] R. Marritz, 'Investing in Efficiency', *Electricity Journal*, Vol. 1, No. 2, August/September 1988, pp. 22–35.
[29] Frederick T. Sparrow et al, 'Equity, Efficiency, and Effectiveness in DSM Rate Design', *Electricity Journal*, May 1992, pp. 25–33.

Are negawatts cost-effective?

This is a controversial and hotly contested question. One recent study by Joskow and Marron, possibly the best to date in addressing this particular question, concludes that:[30]

- No one *really* knows the answer because of the differences in cost accounting used by various entities.

- The answers that are reported by utilities vary sometimes by an order of magnitude depending on the market segment and the types of programmes cited.

Joskow and Marron make the following suggestions to improve the quality of the DSM cost measurement and reporting procedures:

- Better accounting for 'missing' utility costs such as overhead, administration, evaluation, and measurement costs.

- Better accounting for 'missing' customer costs including customers' out-of-pocket expenses to participate in utility DSM programmes.

- Reliance on *ex post* measurement of actual savings (as opposed to *ex ante* engineering estimates of the savings).

- More consistent accounting of the 'economic' life of DSM measures.

- Better accounting of 'free riders'.

- Standardized methods of cost accounting to make the intra-utility results more compatible.

But in fairness, it must be noted that the DSM industry is still in its infancy and much of the disagreement about the cost-effectiveness of DSM programmes is simply disagreement on what should, and should not, be included in the costs, and how the results should be reported.

---

[30] Paul L. Joskow and David B. Marron, 'What Does a Negawatt Really Cost?', *Energy Journal*, Vol. 14, No. 4, 1992. See also Box overleaf.

What does a negawatt really cost? Answers by Joskow and Marron[a]

Joskow and Marron used information reported by ten utilities about their electricity conservation programmes to calculate the life-cycle cost per kWh saved – the cost of a 'negawatt' – associated with these programmes.[b] These computations indicate that the cost associated with utilities 'purchasing' negawatthours is substantially higher than is implied by sources such as Amory Lovins (of the Rocky Mountain Institute) and EPRI.[c] The costs calculated for residential programmes, in particular, were found to be much higher than conservation advocates have suggested. However, 80 per cent of the expected savings from these programmes are attributed to commercial and industrial customers rather than to the residential customers the programmes were designed for.

In answering the question 'what does a negawatt really cost?', the authors are frank and blunt: 'neither we [the authors] nor anyone else really knows with any precision!'. But the basic conclusions of the study may be summarized as follows:

• The negawatt costs – as measured and reported by individual utilities – vary considerably both among different utilities and among different programmes within a single utility, as shown in Table 4. For example, utility 1's residential DSM programmes achieved kilowatthour savings in the range of 3.4 to 21.6¢/kWh, with an average cost of 7.6¢/kWh. For utility 5A, the range was 5.2–181.4¢/kWh, with an average of 22.1¢/kWh. The average US retail cost of electricity for residential customers is 8.1¢/kWh, and for large commercial/industrial customers it is 5.1¢/kWh.

• In examining the utility cost records, the authors have identified several areas with '*significant*' downward bias. The most important are reliance on *ex ante* engineering estimates of savings as opposed to *ex post* measurement; inconsistencies in estimating the life of DSM measures; and failure to account for the so-called 'free riders'. The combination of these and other systematic biases, according to the authors, means that the real costs of negawatthours could be substantially higher than those reported in Table 4, '*by a factor of at least two*', i.e. multiply the numbers in Table 4 by two or more to get the real costs, on 'average'.

According to the authors:

Accurate measurement of energy savings is likely to increase costs by about 50 per cent. Fully accounting for all administrative costs is likely to increase the cost per kWh saved by 10 to 20 per cent. Fully accounting for customer contributions and taking free riders into account is likely to increase costs by about 25 to 50 per cent. In addition, significant

uncertainties remain due to the large variance in assumed measure liveseconomic lives of DSM measures.

What do the authors have to say about the well-publicized, low-cost projections claimed by conservation gurus such as Amory Lovins – who has stated that fully 75 per cent of the electric end-use energy in the United States could be conserved at an average cost of 0.6¢/kWh – before taking account of the savings in electricity costs? The authors write:

We do know that electricity conservation 'purchased' by utilities *costs substantially more* than suggested by estimates of technical potential such as those produced by *RMI, EPRI,* or the *NES* [National Energy Strategy], Department of Energy, has a *higher variance* in costs than implied by their analyses, and is likely to have actual costs that are *significantly understated* in the reports issued by most of the utilities whose programmes we have examined. The disparity between the technical potential estimates and the cost estimates reported by utilities can be traced to the *excessively optimistic assumptions* that underlie the technical potential studies, to the *administrative costs* that utilities must incur in providing conservation services, to the enormous diversity among consumers in current and expected energy utilization patterns, and to the fact that, in a world of imperfect information, utility programs necessarily attract some participants for whom the conservation investments are not cost-effective. [emphasis added]

Notes:
[a] See footnote 30.
[b] The MIT economists used the reported data from the following US utilities with active DSM programmes as the basis for their study (they do not, however, attribute cost numbers to the individual utilities for obvious reasons). The sample utilities had an aggregate DSM budget of over $700 million in 1991:

- Boston Edison
- Central Maine Power
- North East Utilities
- Western Massachusetts Electric
- Long Island Lighting
- New England Electric System
- General Public Utility Corporation

- Pennsylvania Electric
- New York State Electric and Gas
- Northern States Power
- Pacific Gas and Electric Company
- Puget Sound Power and Light
- Southern California Edison
- Wisconsin Electric Power

However, reasonably consistent and useable data were received from only 10 of the above, and are reported. For some utilities, the data from different affiliates were substantially different and are reported as such.
[c] A.P. Fickett, C.W. Gellings, A.B. Lovins, 'Efficient use of electricity', *Scientific American*, Vol. 263, No. 3, 1990, pp. 64–74.

Table 4 Cost per kWh saved by utility programmes 1991, cent/kWh

Utility programme	Residential		Commercial/industrial		Weighted average	All utility costs included?	Direct customer costs included?	Measured savings estimates?
	Average	Range	Average	Range				
1.0	7.6	3.4–21.6	6.7	3.0–10.3	6.9	Most	No	No
2	4.9	2.0–22.3	3.5	1.0–17.1	3.8	No	No	Some
3	10.4	3.0–15.4	1.9	0.8–8.0	2.8	Yes	Yes	Some
4	4.7	0.6–33.3		1.7–5.9	3.3	No	Yes	No
5A	22.1	5.2–181.4	3.1	2.2–9.6		No	No	Yes
5B	6.8	3.9–9.2	4.4	3.7–4.6	4.8	No	No	No
5B1	7.6	3.9–20.2	6.4	3.1–7.3	6.6	No	No	Yes
6	12.4	2.9–14.1	1.5	0.2–4.7	1.9	No	Yes	No
7	4.8	0.4–68.8	1.9	1.3–4.9	2.2	No	No	No
8A	7.2	3.2–22.1	2.4	1.4–18.1	3.0	No	No	No
8B	4.4	2.1–12.6	3.6	2.6–5.3	3.7	No	No	No
9	7.2	3.8–160.6	2.4	1.3–18.4	3.4	Most	Yes	No
10	3.5	0.3–5.5	2.0	1.9–2.4	3.0	Yes	No	No
EPRI	3.0	0.9–3.6	2.3	0.7–3.9	2.5	No	Yes	No
RMI	>2.0	−1.8–4.0	<0.6	−1.8 −1.0	0.6	No	Yes	No
NES	2.9	2.0–3.9				No	Yes	No
Av. retail price*	8.1		5.1		6.9			

Notes: *1990; EPRI=Electric Power Research Institute; RMI=Rocky Mountain Institute; NES=National Energy Strategy, US Department of Energy.

Source: Joskow & Marron, *The Energy Journal*, Vol. 14, No. 4, 1992, pp. 41–74.

What will be the long-term impact of DSM on load growth?

The answer depends on several critical assumptions about the cost and penetration level of DSM programmes in various market segments; government mandated efficiency standards which will remove some inefficient options from the market over time; the overall rate of economic growth which underlies demand for energy services; cost of energy, and a number of other parameters. These are not trivial issues.

Another complicating factor further confusing the debate is whether one is talking about maximum technical potential (MTP) or achievable potential of DSM.[31] One gets entirely different answers depending on how the question is posed. Most estimates start with MTP, and then scale it down by what can realistically be expected over a period of time based on assumptions regarding customer acceptance of new technologies, take-up rates, economics of capital investments versus lower operating costs, and so on. This is tricky business.

Policy questions

Ultimately, the question is not *whether* and *how much* more energy efficiency can comfortably be squeezed out of the US economy, but *how soon* and *how best* to achieve it. Would it be more efficient to continue or accelerate the current regulatory driven approach of setting new standards and to reward utilities for acting as catalysts of change; or would it be better to raise energy prices (perhaps through energy or carbon taxes) to accelerate the penetration of more efficient technology?[32] Would it be better to require the utilities to do most of the DSM programme implementation; or should some or all of it be done through competitive bidding or by energy service companies (ESCOs)?

Is it efficient to rely on utilities to finance and implement what is socially desirable, or should another mechanism be found for accomplishing the same objective? One ex-regulator, for example, has proposed the concept of a *conservation utility* or agency whose sole purpose would be to sell

[31] A. Faruqui et al., 'Efficiency Electricity Use: Estimates of Maximum Energy Savings', CU-6746, Electric Power Research Institute, Palo Alto, CA, 1990.
[32] Arthur H. Rosenfeld and Lynn Price, *Incentives for Efficient Use of Energy: High Prices Worked Wonders from 1973 Through 1985. What are Today's Alternatives to High Prices?*, Lawrence Berkeley Laboratory, Berkeley, CA, June 1992.

'negawatthours', presumably in direct competition with utilities which sell megawatthours.[33] Another ex-regulator proposed charging customers for negawatts as well as kWhs.[34] And if we decide that utilities are the best available agents for the job, then the question becomes: 'What is the best way to motivate them to do what's good for the society?' The preceding discussion clearly demonstrates that the best way to get utilities' attention is through schemes which reward them for doing the right things. But even if we agree on this point, there is still the question of *how much* reward is necessary and how it should be administered.[35]

Incentives

Currently, 25 US jurisdictions offer one or more kinds of DSM incentives (recall Table 1). DSM incentives are the subject of an intensifying debate among regulators, utilities and consumers alike.[36] Proponents advocate them as a means of stimulating expanded utility involvement in conservation and load management. Opponents reject them because they usually increase average rates, at least in the short term, and because non-participants have no veto power over what the utility does, how it selects and implements its DSM programmes, and which customers it recruits for participating in the programmes. This is seen by some critics as a form of taxation without representation. From the shareholder's point of view, however, incentives provide an important offset to the negative impact of declining sales resulting from DSM.

---

[33] D. Nichols and P.D. Raskin, 'Conservation Utilities: New Forces on the Demand Side', *Electricity Journal*, Vol. 2, No. 8, October 1989, pp. 18–25.

[34] M.B. Katz, 'Utility Conservation Incentives: Everyone Wins', *Electricity Journal*, Vol. 2, No. 8, October 1989, pp. 26–35.

[35] Richard Gilbert and Steven Stoft, *Incentives for Utility DSM: A Classification and Analysis*, Energy Research Group, University of California, Berkeley, June 1992; Michael W. Reid and John H. Chamberlin, 'Financial Incentives for DSM Programs: A Review and Analysis of Three Mechanisms', American Council for Energy Efficient Economy (ACEEE), 1990.

[36] M. Reid, 'The Evolution of DSM Incentives', in S. Nadel, M. Reid and D. Wolcott (eds), *Regulatory Incentives for Demand-Side Management*, ACEEE, Berkeley, CA, 1992.

Lessons from the US experience

Although the jury is still out on this, some conclusions can be drawn from the US experience of DSM so far:

- DSM incentives have generally been effective in motivating investor-owned utilities to take them seriously.

- Even though much has been learned, shared and published over the past several years on DSM, many unresolved issues and questions still remain.

- Many trend-setting utilities are now relying on DSM for a growing share of their future energy and capacity needs.

The regulatory zeal to push DSM (and IRP) over a short period has created a few problems and has raised a number of persistent and irritating questions for which there are no easy, generally accepted, answers as yet. But with the level of energy and resources going into research, it won't be too long before some, if not all, of these issues are resolved. The following is a sample:

- *High profits?* Some critics of DSM argue that utilities should not be allowed high profits on their DSM investments.

- *Evaluation problems* There have been allegations that some utilities may be collecting too much money from ratepayers on unsubstantiated and exaggerated DSM programme results. Regulators are certainly aware of these problems and are tightening the rules.

- *Persistence* How long do the benefits of DSM programmes last? We know that there are some 'snap-back' effects following the introduction of certain measures, but there is limited empirical evidence on longer-term persistence of DSM benefits.

- *Cost effective DSM?* This issue, as already alluded to, is a controversial one.

- *Higher rates?* So far, the impact of DSM programmes on utility rates has been relatively small. But as more and more utilities rely on DSM for a growing share of their future resources, this issue needs to be re-examined.

- *Is it fair?* The problem of non-participants getting the short end of the DSM deal is a sensitive issue. Ultimately, policymakers must decide (and many have already decided) that some things do more good for society as a whole than harm to some sectors of society, and are therefore worth pursuing.

- *Industrial customers* Large customers continue to complain that non-involuntary payment for utility-sponsored DSM programmes is unfair. They also dislike subsidizing their competitors for things they should have done in the first place and out of their own pockets. Some jurisdictions are sympathetic to these allegations and are considering schemes that will allow large customers to declare themselves 'ineligible' for DSM programmes – hence no longer obliged to pay the higher rates.[37] But if you let some customers go, what do you say to others who may wish to follow? Where should one draw the line?

- *Customer sovereignty?* Carrying the previous argument one step further, shouldn't we let all customers decide whether they wish to pay for utility-sponsored DSM programmes? After all, why shouldn't those who directly benefit from DSM programmes pay for the costs?

Policy implications

Given these unresolved issues, what can the overseas observers learn from the US experience of DSM to date? And how much of the US experience is applicable to the rest of the OECD, eastern Europe, or the less-developed countries (LDCs)? Here are some observations:

- *Regulations work, but up to a point* The US experience suggests that while there is ample room for certain types of command-and-control regulations/standards at the state and federal level, there are limits to what they can achieve. For example, housing construction codes, appliance or automobile efficiency standards or labels are a cost-effective way to gradually cleanse the market from clearly inefficient and inferior products,

---

[37] *Demand Side Monthly*, October 1992.

but it is not easy to encourage customers to replace their existing appliances with super-efficient ones until the old ones wear out. Rebates and incentives can go a long way to accelerate the turnover rate of old appliances and increase the market penetration of newer, more efficient ones.

- *High energy prices/taxes motivate conservation* Energy conservation is second nature in countries with high energy prices. No incentives are needed to sell energy-efficient appliances or automobiles because the rewards are real and automatic. When energy prices are stable and low, as they have been in the US since 1981, it takes much more effort to sell energy conservation. Proponents of high energy taxes point to such benefits as partial justification for raising energy taxes.

- *DSM incentives work* The experience of the past several years in the United States clearly shows that the profit motive is a strong inducement. Other motivations (e.g. environmental benefits of DSM) also help. The level of incentives, however, has to be set out in a way that is not perceived as 'over generous'.[38]

- *Good implementation is the key to success* Utilities that have done well in their DSM programmes have put ample resources into their planning, implementation, and evaluation.

- *Better accounting is necessary* Regulators setting the bait (in the form of DSM incentives) need to examine the reported savings. The more attractive the rewards, the greater the likelihood of exaggerated claims.

- *Uniform standards are a must* The surprising lack of uniformity of results reported in studies such as that of Joskow and Marron highlights the need for basic accounting and reporting standards. Work is already underway in this area.[39]

- *Learning by doing* really works when it comes to DSM.

---

[38] Some utilities have been earning 25 per cent interest on their DSM investments, which is twice what they make on their other investments.

[39] For example, Eric Hirst and Carol Sabo, 'Electric-Utility DSM Programs, 1990 to 2010', ORNL/CON-312, Oak Ridge National Laboratory, Oak Ridge, TN, 1991.

PART TWO: PROMOTING RENEWABLE ENERGY

Background

The motivations behind the push to promote renewable and alternative (R/A) energy in the 1970s were almost identical to those for pushing energy conservation and DSM.[40] And, once again, there were efforts at the federal as well as the state level.

A key element in the federal government's plan to commercialize renewable and alternative technologies (R/As) was to guarantee a market for the generated electric power at an attractive price. This was provided by the passage of the Public Utility Regulatory Policies Act of 1978, better known as PURPA. Under PURPA, utilities were required to buy all that was produced by Qualifying Facilities or QFs[41] and were required to pay for QF power based on the utilities' 'avoided costs'.[42] Utilities were also required to interconnect with such producers and provide supplemental and back-up power to them at 'fair and reasonable' rates.

The logic behind PURPA was fourfold: it was supposed to reduce the nation's dependence on conventional fossil fuels by encouraging the development of R/As; it was expected to lead to competition in power generation – hence stabilizing or reducing electricity costs; it was intended to diversify the energy

---

[40] Throughout this paper a distinction is made between 'R/As' – which includes renewables and alternative power (including cogeneration and coal gasification, etc.) – and 'renewables', which are non-fossil based and renewable in nature.

[41] As defined by PURPA, a QF is either a 'small power production facility' that produces power from biomass, waste products, geothermal, or renewable resources and has a maximum capacity of 80MW, or is a 'cogeneration facility' that sequentially produces electricity and useful thermal energy and meets certain minimum efficiency and performance standards. A cogeneration facility may use conventional or renewable fuels and is not subject to any size limitation. In either case, to be regarded as a qualifying facility, such a plant may not be more than 50 per cent utility-owned.

[42] The price that utilities were required to pay for QF power was stipulated as the utilities' 'avoided costs', defined as the anticipated marginal or incremental costs to the relevant utility for energy and/or capacity that the utility would have incurred if it were to build a new plant or acquire power through other means. QFs were offered attractively high rates because the marginal costs usually exceeded average utility costs and also because in the late 1970s and early 1980s the price of oil (as well as of natural gas) was high and expected to rise.

mix of the electric utility industry; and it was to accelerate the commercialization and deployment pace of R/As. As will be documented below, PURPA has been highly successful in achieving some of its intended goals, and partially successful in other areas.

The real impetus for oil displacement and energy conservation, however, followed the 1979–80 Iranian revolution and oil price shocks, which saw spot prices for oil top $44 per barrel in 1980. More importantly, in the ensuing panic, many industry analysts estimated that oil prices would continue to rise, reaching $100 per barrel by the turn of the century.

These developments combined to produce a highly charged, over optimistic outlook for the R/As. The main reasons for this euphoric atmosphere were:

- Oil prices were high and were expected to rise even further.

- Generous federal and state tax credits – designed to encourage development – had the effect of reducing the apparent cost of R/As.

- Federal (and state) funding for R/A research, development and demonstration (RD&D) projects was plentiful and rising.

- Researchers and R/A proponents were projecting rapidly falling capital and operating costs for R/A technologies, assuming fast commercialization and mass production of components.

- PURPA had created a guaranteed market for the generated power at highly attractive prices at essentially zero risk.

- With a new energy conservation ethic, the country was receptive to 'small is beautiful', and 'dispersed is better than centralized'.

Given this combination of factors, it is not difficult to see the appeal of predictions that R/As would become cost competitive with oil-based generation in the late 1980s or early 1990s.

Southern California Edison Company (SCE) took a bold initiative in October 1980 to aggressively accelerate the development and commercialization pace of R/As.[43] SCE's corporate goal was to meet at least one-third of its new capacity needs for the next decade – more than 2,000MW – from R/As. This

---

[43] Sioshansi and Nola, 1990, op. cit.

effort, which was strongly endorsed by the California Public Utilities Commission (CPUC), was driven by four primary considerations:

- *Financial stability* R/As were viewed as one way to encourage external financing of capital investments for new generation between 1980 and 1990. SCE's expectations at the time were that at least 50 per cent of its R/As capacity goals would be externally built and financed. As it turned out, external financing accounted for a much higher percentage.

- *Rate stabilization* SCE expected R/A development to help stabilize its rates, which were projected to rise because of its ongoing nuclear construction programme. In retrospect, the reverse happened.

- *Environmental benefits* Another component of SCE strategy was to protect the environment by relying on more benign, environmentally responsible R/A technologies. This goal has been partially achieved.

- *Energy diversity* Perhaps the most important objective of SCE's goal was to reduce its dependence on foreign oil supplies and to diversify its energy mix. As will be described below, this objective has also been partially met.

In pursuing these objectives, SCE embarked on an all-out RD&D effort that catapulted it to the forefront of R/A commercialization worldwide. During the first half of the 1980s, with the financial support of the Department of Energy, the Electric Power Research Institute (EPRI), and a number of other participants, SCE built and operated one of the world's largest experimental windmills, the world's largest solar central receiver facility, the world's first and largest coal gasification combined-cycle facility, and two of the largest geothermal power plants in the United States utilizing hot geothermal brine. SCE also experimented with a variety of other exotic R/A technologies developed by private investors. With the implementation of PURPA, SCE became the first utility to use nine different primary energy sources on its system, more than any other utility in the world.

During the same period, and with the full support and encouragement of the CPUC and the California Energy Commission (CEC), SCE took specific steps to make it easier for R/A developers to obtain financing for R/A projects. SCE (under pressure from the regulators) was generous in its early contracting practices by offering attractive floor or levelized payments above avoided cost for R/A facilities. In exchange for early project price support, SCE was willing to take discounts on avoided cost payments in later contract years such that, on an overall basis, the purchased cost to the ratepayers would be no greater than the avoided cost over the life of the contract. In addition to pricing support, SCE assisted the early R/A developers with feasibility studies, interconnection support (with cost recovery through discounts on energy and capacity prices), and even leasing of acquired land in the case of wind park developers.

CPUC, in its efforts to encourage the development of R/A resources, required the California utilities to offer Standard Offer (SO) power purchase contracts based on avoided cost principles. The early SOs, however, did not provide any form of price certainty over the life of the contracts. The CPUC also encouraged utilities and developers to negotiate 'non-standard' contracts, since developers were not required to take SOs unless they wished to. The CPUC was so insistent on the development of R/A resources in California that it penalized both SCE and Pacific Gas & Electric Company (PG&E) several million dollars in their general rate cases for allegedly 'not doing enough' to accommodate R/A developers during the early 1980s.

By 1983, the R/A industry in California had grown strong enough to influence the CPUC to establish a Standard Offer which provided more price certainty during the early years of the project than was previously available. The CPUC approved these new SO energy payment provisions based on escalating marginal fuel price projections for up to ten years of a 30-year contract. All of these efforts, of course, were to make it possible for the fledgling R/A industry to get the necessary financing and invest in new technologies. In retrospect, the CPUC's overly aggressive R/A policy proved too successful, and the situation got out of hand. And when the time arrived for the brakes to be applied to slow down the speeding train, the mechanisms to do so – and to do so in a timely manner – were not there.

Unexpected turnaround

Two dramatic developments have taken place on the energy scene since the early 1980s. The most significant turnaround has been the persistent drop in oil prices experienced since 1981. Not only have the prices fallen, but successive projections of future oil prices have been scaled down to what amounts to a flat curve (in real terms) for the remainder of this century, not reaching the equivalent in real terms of their all-time high of more than $44 per barrel until well into the next century. This represents a dramatic change from the widely accepted expectations of the late 1970s.

The significance of this turnaround cannot be overstated for two reasons. First, since the price of oil is often used as a yardstick against which all other energy prices are measured, falling oil prices have the effect of lowering the price of alternative fuels such as natural gas. Second, since the price of oil has a direct impact on the 'avoided cost' of energy – i.e. the marginal cost of energy, which is oil- or natural gas-fired combustion turbines for most utilities – lower oil prices mean lower avoided costs, making R/As less attractive economically.

Another significant factor is that cost goals and market penetration projections for R/As – which were considered realistic in the late 1970s and early 1980s – have, in fact, turned out to be highly optimistic. As a consequence, current cost projections for some R/As are not nearly as encouraging as they were once believed to be.

The combined effect of substantially lower oil prices and the slower than expected pace of commercialization for R/As is clear: R/A technologies which were expected to become cost-competitive with oil in the late 1980s and early 1990s will not become so until the late 1990s or later. With current oil price projections, some of the more exotic, less developed technologies are not expected to become economically feasible until the turn of the century – if ever. To make matters worse, federal and state interest and funding for energy-related R&D were drastically cut during the Reagan and Bush administrations.[44]

Ironically, concerns about pressing environmental issues such as global warming have led to renewed interest in non-polluting R/A technologies. However, with the exception of wind, renewable energy technologies have lost momentum for the time being in the US, and near-term investment in R/As is projected to be static over the next few years. There are a number of

---

[44] Roland J. Sutherland, 'An Analysis of the US Department of Energy's Civilian R&D Budget', *Energy Journal*, Vol. 10, No. 2, 1989, pp. 35–54.

reasons for this. For example, wind park developers now have to face local opposition to visual impact, noise, and land erosion; biomass and waste-to-energy projects suffer from the 'not in my backyard' syndrome and concerns about carcinogens and air quality; hydro projects are vehemently opposed by conservationists on grounds of damage to the environment; and the cost of solar technologies, such as photovoltaics, has not fallen as far or as fast as expected.

The big PURPA surprise

The impact of PURPA on the US electric power industry has been far more dramatic than, and different from, any predictions which could have been made in 1978. Although opinions vary on whether – on balance – PURPA has been 'good' or 'bad' for the country, the consumers, or the nation's utility industry, some observations can be made on the subject.

Too much, too soon

As seen from the utilities' vantage point, particularly in states such as California, PURPA's pace of development has been too fast, resulting in overcapacity and electric system operational problems such as minimum load situations.[45]

California as a state has nearly 16,000MW of R/A contracts with independent PURPA qualifying producers, of which about 6,000MW is currently on line. With a total state generation capacity of 41,000MW, this represents a sizeable contribution. The breakdown of the installed capacity by fuel type for the three large investor-owned utilities in the state is as shown in Table 5. But because of the marked differences in the capacity factors, the energy generated is far more skewed, with cogeneration – fuelled primarily by natural gas – accounting for the lion's share of output. (See Figure 2.)

---

[45] Since QFs make money primarily by selling kWhs (not kWs), they have strong economic incentives – through the implementation of PURPA by state regulatory commissions – to operate their facilities as base-load units at very high capacity factors. This forces the utility to shut down or reduce to minimum levels its own generation, or reject low-cost import electricity during off-peak hours since the QF power must be taken whenever it is generated.

Table 5 Installed R/A capacity breakdown by type: California investor owned utilities 1988, per cent

	PG&E	SDG&E	SCE	Total
Biomass	10	4	7	7
Cogeneration	52	92	55	55
Geothermal	5	0	12	9
Small hydro	7	3	3	5
Solar	0	0	6	3
Wind	26	1	19	21

Notes: PG&E=Pacific Gas and Electric Company, San Francisco, CA; SDG&E=San Diego Gas and Electric Company, San Diego CA; SCE=Southern California Edison Company, Rosemead CA.
Source: Quarterly Report to the CPUC by PG&E, SDG&E, and SCE.

Figure 2 SCE purchases from non-utilities in 1989, billion kWhrs

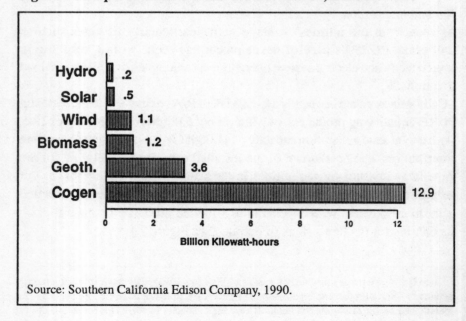

Source: Southern California Edison Company, 1990.

Too expensive

The primary reasons for the rapid development pace of R/As in early years were the generous federal and (in California) state tax incentives (now phased out), the highly attractive 'avoided cost' rates and forecasts of energy costs that were guaranteed in the early SO contracts.[46] Consequently, a flood of applications poured in to the California utilities for contracts that were essentially riskless.[47] The CPUC was too slow in revising contract price forecasts or limiting the availability of these lucrative contracts. For example, SCE purchased more than one-quarter of its total energy requirements and nearly one-half of its fuel and purchased power expenses from QFs in 1989.[48] Due to the fixed cost nature of the Standard Offer contracts, SCE is estimated to 'overpay' for QF power at the rate of $1.3 billion over the lives of these contracts, a burden that will be borne by its customers through higher electricity rates.[49] The situation is much the same at PG&E, the other major investor-owned utility in California.

Fuel diversity?

If fuel diversity was one of the original objectives of PURPA, all indications are that this goal has not been totally achieved. An overwhelming percentage of PURPA-generated energy comes from cogeneration, mostly natural gas-fired. Renewable fuels such as wind, solar, geothermal, and small hydro account for a small percentage of the total kWhs generated. The net result is that electric utilities which receive a lot of PURPA power – such as SCE and PG&E – have, in fact, become *more* dependent on natural gas, even though their internal use of this fuel has fallen. But even with gas-fired cogeneration, it can be argued that their overall process efficiency has improved.

---

[46] 'California will pay dearly for PURPA power', *The Wall Street Journal,* 31 March 1987.
[47] Since both the market was guaranteed and the prices were fixed, R/A developers had essentially guaranteed income revenues for the term of their contracts (or at least the first ten years), if they were able to perform and generate electricity. In California, utilities were required to sign any properly filled-out SO tendered to them by a PURPA qualifying developer.
[48] *Southern California Edison (SCE) 1989 Annual Report and Statistical Summary*, SCE, Rosemead, CA, 1989.
[49] This estimate is based on the difference between the cost of power that could have been purchased from other low-cost producers and/or SCE's own base-load facilities and those paid to QFs under existing contracts.

These criticisms notwithstanding, one can argue that PURPA has resulted in substantial increases in renewable power generation. According to one recent estimate, there is currently more than 5,100MW of installed renewable capacity in the United States, two-and-a-half times the amount in 1984.[50] Another 6,100MW is expected to be installed by 1995.[51]

Hence, the fuel diversity impact of PURPA is a matter of interpretation. But if one takes away the contribution of geothermal and biomass, other renewable resources have not had quite the impact they were expected to make. This is particularly true since wind and solar technologies tend to have fairly low capacity factors.

Is small beautiful?
One of the favourite arguments of the proponents of R/As was that small, modular, dispersed power generation technologies are inherently better (i.e. safer, more robust, less vulnerable, etc.). In practice, however, this argument breaks down in two areas. First, looking at the kWhs (not kW), it is clear that a large portion of the PURPA energy production is generated by a small number of large cogeneration facilities operating at high capacity factors. These facilities do not necessarily fit the mode of small is beautiful.

Second, since renewable energy – with the possible exception of geothermal and biomass – tends to be highly diffused, R/As are highly land- and capital-intensive. Wind and solar technologies, by nature, take a great deal of land and require significant amounts of capital for producing an equivalent unit of reliable (i.e. firm) capacity. In this sense, even though individual wind machines may seem benign and unobtrusive, massive wind parks – consisting of hundreds of machines – are highly visible and conspicuous. Furthermore, even though wind energy production displaces fossil-based energy sources, its availability does not necessarily match the utility's peak period requirements (depending on location). Hence, wind power does not always reduce the need for building additional peaking capacity.

---

[50] California accounts for more than one-half of the renewable total.
[51] 'In pursuit of power', *The Economist*, 28 October 1989.

Operational and integration problems

From the point of view of routine operations and integration into an existing utility grid, PURPA power has been a logistical headache because the number, size, location, timing, and operations of QFs do not necessarily match the utility's system requirements. As in the case of California, too much QF power can come on line for reasons that are beyond the control or need of the utilities which will be the ultimate recipients of the energy and capacity. This gets to be a problem when the scale of QF power reaches the magnitude that it has achieved in states such as California and Texas.

From an operational standpoint, the day-to-day and hour-to-hour coordination and management of QF power also creates problems. Unless the utility has the ability – physically and contractually – to control QF generation, it becomes the tail that wags the dog – i.e. the utility has to adjust its internal operations to accommodate whatever the QFs are producing. Much has been learned about these integration issues, but more needs to be done in the area of QF dispatchability and control.

Contractual problems

Another set of problems arises in negotiating QF contracts of different size, fuel type, capacity factor, dispatchability, and so on. Invariably, a sticking point in all negotiations is the determination of the value of QF power depending on its availability, dispatchability, capacity factor, and other variables. As already mentioned, too much 'uncontrolled' QF energy can create serious minimum load problems for the utility. The question is, in the course of negotiating and enforcing hundreds of contracts, how should the utility safeguard the interests of its ratepayers and shareholders while carrying out the federal and state mandates of regulators? These are controversial issues.

Lessons learned – the hard way

Since California accounts for roughly one-half of all R/A development in the United States, the experience of California utilities in the development of R/A energy resources should serve as an example for other states and countries contemplating QF-type experiments. The first lesson is to examine and – to the extent possible – anticipate the ramifications of new policies or regulations before they are implemented. PURPA provides classic examples of unanticipated (but perhaps predictable) and unintended outcomes. It also provides examples of how resourceful entrepreneurs can outguess and outrun the best intentions of the policymakers. The second lesson is to monitor closely the implementation pace of new policies and be prepared to make timely adjustments, as necessary.

Other hard-learned lessons of the California experience with PURPA can be summarized as follows:

- Do not require utilities to pay for capacity that is not required or in advance of need (i.e. put limits on capacity needs based on each region or utility's projected requirements, and only pay for what is needed to meet customers' demand).

- If SOs are to be used for contracting with independent power producers, allow those offers to be available only for blocks of capacity consistent with need (i.e. do not leave open the availability of SOs such that an overcapacity situation could develop).

- Do not fix prices on forecasts that cannot be updated or withdrawn (i.e. do not set fixed prices for avoided costs).

- Provide for utility dispatch of QF generation to avoid minimum load and similar operational problems.

- Develop a high level of communication and coordination between QF and utility, both during project development and operation. A smooth working relationship between the two entities will result in the QF operating as if it were a part of utility's own generation resources, benefiting both parties.

In retrospect, the lessons of California's experience with PURPA are not all negative. Indeed, the world has reasons to be grateful that California's electric utility customers effectively subsidized much of the RD&D cost of many renewable energy technologies, particularly wind. This has positively contributed to the commercial deployment of wind energy not only in California, but around the world. If there are any regrets, they concern the efficiency and fairness of this hidden RD&D subsidy.

Re-evaluating the role of R/As for the 1990s

Looking ahead to the decade of the 1990s and beyond, is there a new, more mature role for the R/As to play? And if so, what is the new role of electric utilities? Recent developments suggest that renewed interest in R/As may be warranted. For instance, growing concerns for environmental protection are likely to dominate the US and global energy decisions in the 1990s. Issues ranging from global warming and carbon dioxide build-up in the atmosphere to local issues like improving the air quality in Los Angeles will continue to gain momentum.

The reason for stringent air emission measures is quite simple. Air quality is worsening throughout the United States. In fact, in March 1989, the Environmental Protection Agency reported to Congress that 96 areas exceeded federal limits on low-level ozone – 28 more than in 1987. This translates to roughly 150 million Americans living in communities with 'unhealthful' air. The situation is equally alarming in other industrialized and developing countries.

The growing significance of environmental issues suggests that some R/As which were prematurely dismissed in the mid-1980s on the grounds of high capital and operating costs may be more competitive if environmental concerns and other externalities associated with conventional power generation technologies are taken into account. Given the growing awareness of environmental problems, industrial societies will increasingly be willing to bear greater economic costs for cleaner power generation. Many renewable technologies, such as solar and geothermal, emit no greenhouse gases to the atmosphere.

Past comparative studies of alternative generation options have generally neglected to factor the broader social and environmental benefit of R/As into the analyses. To consider the true value of R/As, it is necessary to include *all* the indirect and intangible social costs and any negative externalities associated with conventional power generation. Any feature of power generation that contributes 'disutility' to members of society, either currently or at some point in the future, can be considered an externality or social cost.[52]

---

[52] Fereidoon P. Sioshansi, 'Demand-Side Management and Environmental Externalities', op. cit.

Externalities associated with power generation include air emissions, consumption of non-renewable resources, land use, water use, solid wastes, hazard risks, and visual impacts. Another important factor to consider is the modularity, i.e. division into smaller units with shorter construction times, of many R/A technologies. The critical question, of course, is how can these externality costs be measured and quantified? This is obviously a major hurdle to overcome. Increasingly, however, data are becoming available that can begin to address these questions.

Another promising note for R/As is that, in the future, utilities across the United States will be utilizing some form of multi-attribute bidding procedures to choose among competing demand-side and generation technologies. These bidding procedures invariably have provisions that attempt to credit R/As for their environmental, societal, and other non-tangible benefits. While it is not clear how much some of these non-quantifiable benefits are worth, all indications are that the concept of giving R/As some credit for these positive attributes is gaining momentum.

Conclusions

PURPA and the experience of renewables of the last decade, particularly as it applies to California, should be required study for any country or regulatory entity contemplating similar policies in the future. Taking everything into account, it has been a mixed blessing. More significantly, it should provide the policymakers with a list of valuable dos and don'ts for implementation – many of which were learned the hard way in California. With hindsight, it is easy to criticize some of the implementation oversights of PURPA, but at the same time, one should give PURPA credit for some of its unanticipated successes.

On the positive side, PURPA has encouraged the development of renewables and has resulted in some fuel diversity – but the results have been highly skewed in favour of natural gas-fired cogeneration. Furthermore, PURPA has spawned a burgeoning new QF industry which, by many estimates, will dominate the new generation capacity needs of the US electric power industry into the 1990s and beyond. This has, in effect, removed the electric power companies' monopoly position in generation and – by some accounts – may eventually lead to expanded competition in transmission and distribution of electricity. On these grounds, PURPA has been far more successful than anyone could have anticipated in 1978.

On the negative side, because of excessively aggressive implementation policies, R/A development in certain areas of the United States has resulted in overcapacity situations, minimum load conditions, dispatchability problems, and higher electricity rates than would have otherwise been the case. The affected electric utilities face considerable loss of control in operating their own systems, have had to forgo low-cost power purchase opportunities, and have had to reach compromises with R/A operators in operating and dispatching their facilities. Many more controversial technical and policy issues remain to be resolved.

PURPA demonstrates that good regulatory and policy intentions are not sufficient. In developing well-structured public policy, careful attention must be given to implementation issues.

Part 4

Opportunities and Strategies

The Economics of Energy Savings: CO_2 Emissions, Energy Consumption and Competitiveness of Industry in France

Françoise Garcia, Head, Economic Department, ADEME (Agency for Environment and Energy Management), Paris

Sustainable Energy Policies: Political Engineering of a Long Lasting Consensus

Ernst von Weizsäcker, President, Wuppertal Institute for Climate, Environment and Energy

The Restructuring of the Electric Utility: Technology Forces, R&D and Sustainability

Carl Weinberg, Weinberg Associates – Sustainable Energy Development, California, formerly Manager, R&D at Pacific Gas and Electric Company, California

Development and Transfer of Sustainable Energy Technologies: Opportunities and Impediments

Katsuo Seiki, Executive Director
and
Yutaka Tsuchida, Manager, Global Industrial and Social Progress Research Institute, Tokyo

The Economics of Energy Savings: CO_2 Emissions, Energy Consumption and Competitiveness of Industry in France

Françoise Garcia

Abstract

The lack of consensus at EU or OECD level on the levying of carbon or other energy taxes has prevented the adoption and implementation of a significant common 'price signal' for CO_2 reduction.

In the present environment, it is difficult to implement tax instruments at a level which would help to prevent global warming. Concerns about competitiveness remain a crucial factor in preventing a wide consensus. The lack of price incentive and the economic crisis further hinder the commitment of industrial firms to energy conservation. This paper draws on the results of two studies, carried out for Agence de l'environnement et de la maîtrise de l'énergie (ADEME) and the French Ministry of the Environment, which had the aim of investigating the particular opportunities for and constraints on energy conservation in the industry sector. Results indicate that:

- *There is an estimated conservation potential of 19 per cent of demand in the early 1990s in French energy consuming industries; most of these efficiency measures have on average a pay-back period of 2–4 years.*

- *Both process and non-process industry operations are important.*

- *There is scope for increased efficiency both in electricity use and in direct fuel use.*

- *Only about one-third of the conservation potential would be taken up in the 'business-as-usual' scenarios.*

- *The EC energy/carbon tax as proposed in 1993 is not sufficient to change greatly the technologies that industries use. Instead, as industries passed on the cost of the tax in higher prices, major energy savings would come from downstream substitution between cheaper products and materials. There would also be geographical shifts as firms moved heavily taxable activities out of the tax area, raising concerns about impacts on French industry.*

However, a related study noted that there would be scope for sector voluntary agreements to be made. These would allow sectors and government to discuss what could and/or should be achieved and by what methods. Formal legal requirements might then not be necessary. However, it seems that the potential of contracts for curbing emissions is much more restricted than that of a tax on energy. On the positive side, they are free of macro economic effects due to decreasing competitiveness which is inherent in the tax. Contracts can be more easily reversible and as such are a 'no regrets' instrument, but their impact would be modest.

Energy is at the heart of many environmental challenges, and the concept of sustainable development relies on rational management of our modes of energy consumption and production in the long run. It is recognized that, despite the remaining scientific controversies, action has to be taken now to curb greenhouse gas emissions. Because of the uncertainties a 'no regrets' approach prevails which focuses on low-cost energy-efficient measures.

Among industrialized countries, France has one of the lowest level of CO_2 emissions per capita, about 1.9 tonnes of carbon equivalent per capita per year. This performance results from two aspects of energy policy: energy conservation and nuclear electricity generation. CO_2 emissions in 1991 were 142 gigatonnes (Gt) lower that in 1973. Of this reduction, one-third can be explained by energy savings and two-thirds by energy substitutions mainly due to nuclear power.

In France, as in many industrialized countries, over the period 1973–92, a huge improvement in energy efficiency occurred, most rapidly in the years of high energy prices and strong conservation policy and more slowly in the rest of the period. Energy intensity decreased by 36 per cent between 1973 and 1991. The increase of the energy intensity in 1992 and 1993 has been a cause for some concern.

In 1992, compared with 1973, a total of 28 million tonnes of oil equivalent (Mtoe) of final energy consumption has been saved: 7.7Mtoe in industry, 14.6Mtoe in residential and commercial use, and 6Mtoe in transportation. Compared with the peak of energy conservation in 1989, in 1992 3.2Mtoe of increased consumption appears to be due to behaviour changes – resulting in more energy being used – particularly in the transport and residential sectors, and possibly as an effect of low levels of use of production capacity in industry.

In 1992, industry (steel included) accounted for 54Mtoe and 28.8 per cent of the total energy consumption (compared with 38.2 per cent in 1973). The contribution of industrial and agricultural energy consumption to CO_2 emissions in 1991 was 14.5 per cent from a total of 389Gt of CO_2 in 1991 compared with 21.7 per cent of 531Gt in 1973.

France signed the Rio Climate Change Convention in 1992 and presented its national programme for global warming prevention to the EC in 1993.

The report published by the Government Interdepartmental Committee on the Greenhouse Effect under the chairmanship of Yves Martin came out in favour of taxation as the best means of combating the greenhouse effect. It upholds the introduction of an incremental taxation programme over a wide geographical area covering at least the OECD zone. To enable the economy to adapt, the rate of taxation should grow only slowly but the report stressed that the aim of stabilizing CO_2 concentrations requires the level to reach around FF2000 per tonne of carbon.

Professional organizations and public authorities have criticized the EU plan of action, which includes a tax on energy, based 50 per cent on carbon content and 50 per cent on energy content, with a gradual rise to reach the equivalent of $10 per barrel of oil equivalent (approximately FF500/toe) or about FF450 per tonne of carbon. On the one hand, the impact on international competitiveness is crucial; on the other hand, the mix of energy and carbon is a concern insofar as it could lead to inconsistencies of CO_2 reduction, especially in the case of France. The industry sector – in particular those parts of it developing intermediate materials and goods, which are in most cases exposed to international competition – appears to be quite vulnerable to the impact of such tax action.

As yet no consensus has been obtained at EU or OECD level preventing the adoption and implementation of a significant common 'price signal' for CO_2 reduction. Several countries have increased their energy taxes, partly to raise revenues, partly to demonstrate their commitment to this issue. In France, in July 1993, the domestic tax on fossil fuels was increased by a small amount for industrial uses and by a higher amount for motor fuels. The increase of 28 centimes per litre on motor fuels is equal to 80 per cent of the FF500/toe; the increase on domestic fuel oil was 3.6 centimes per litre, or about 10 per cent of the FF500/toe tax; the increase on heavy fuel oil was only FF10/toe.

In the present situation, it is difficult to make significant use of tax instruments to prevent global warming. Concerns about competitiveness remain a crucial factor in preventing the achievement of a consensus.

Energy conservation is widely recognized as a key element and opportunity in responding to climate change and energy concerns. But the lack of price incentives – and the economic crisis – hinders the commitment of firms to energy conservation. Hence two studies were carried out to investigate the particular opportunities and constraints for the industry sector:

- *An assessment of the potential for energy conservation in French industry from 1990 to 2005*. This study was carried out by a consultancy group, the Centre d'études et recherche sur l'énergie (CEREN), for ADEME. It relies on a technical-economic analysis with an inventory of 188 energy conservation opportunities at a disaggregated level of 23 industry sectors and 18 energy uses. Information was collected from industrial firms, technical centres, professional trades, energy utilities and ADEME.

- *A survey focusing on the CO_2 taxation issue and the impact on the competitiveness of French industry*. A statistical analysis was carried out by the research group Centre d'études sur les ressources naturelles (CERNA)[1] for ADEME and the French Ministry of Environment of 600 industrial sub-sectors to analyse the sub-sector vulnerability. Next a survey was made of 22 major industrial groups representing the most heavily taxed branches. It investigated the effect of the tax on their ability to compete, and the possible energy savings and substitutions, geographical movement of activity and slimming down in activities which the tax might bring about. Third, a comparison was made between the CO_2 energy tax and sector by sector contracts. The final step was a more searching analysis of various instruments and their impact on the implementation of potential energy conservation.

The CEREN survey

In industrial firms facing tough competition, energy management is considered to be efficient in most cases, especially in activities in which energy costs form a large part of production costs. This – often well justified – coupled with the improvement in energy efficiency achieved in the 1970s and 1980s can make policymakers and industrialists suspicious of new cost-effective opportunities.

The study undertaken by CEREN for ADEME takes into account a scenario of economic growth and energy price changes. CEREN provides a forecast of energy conservation up to 2005.

---

[1] The official title is CERNA – Centre d'Économie Industrielle; it is attached to l'Ecole des Mines de Paris.

In the early 1990s, 19 per cent of energy consumption and 9.4Mtoe could be saved if identified available cost-effective techniques were implemented. Of the potential savings, 3.2Mtoe lie in electricity consumption and 5.6Mtoe in direct fossil fuel use. These figures represent 13 per cent and 18 per cent respectively of 1990 consumption. A further 0.6Mtoe could be saved by using cogeneration. (See Table 1.)

Improved information about the potential for electricity conservation could bring electricity savings closer to those of fossil fuel. In the early 1990s more emphasis has been put on specific electricity uses such as motors and lighting.

Over time the progress of research, design and development (RD&D) will enlarge the field of available technologies. This will result in a further potential saving of 1.5Mtoe, mostly taken up by 1995 due to an inability to project or estimate further developments. This amount of 1.5Mtoe, called 'additional RD&D potential', cannot be considered as the entire impact of RD&D because some has, of course, already been included in the 1990 potential. So by 2005, adding the further 1.5Mtoe from identified technology development to 9.4Mtoe, the energy conservation potential reaches almost 11Mtoe – that is, 22 per cent of today's energy consumption.

The analysis is based on the low energy price scenario and the low economic growth scenario defined in the 'Energy forecast up to 2005', a report carried out by the French Energy Planning Commission, assuming figures of US$21 per barrel of imported fuel oil, economic growth at 1.6 per cent annually and annual industrial growth at 1 per cent.[2]

Taking into account the available conservation potential in 1990 of 9.4Mtoe; the impact of economic growth from 1990 to 2005 on potential conservation of 0.6Mtoe; the added RD&D potential of 1.5Mtoe; and a business-as-usual trend of energy efficiency investment in the period of 3.9Mtoe, the potential still available in 2005 amounts to 7.6Mtoe (called 'residual potential'). (See Table 2.)

Using the above assumptions, 36 per cent of the energy efficiency potential will be implemented by 2005, leaving a huge part of it unexploited. This

---

[2] *Perspectives énergétiques de la France a l'horizon 2010*, Report of the working group 'Perspectives Energétique' du Group de Prospective sur l'Enérgie chaired by M. Michel Pecqueur, Xème Plan 1989-1992, Observatoire de L'Enérgie, Ministère de l'Industrie et du Commerce Extérieur, Paris, 1992.

Table 1 Conservation potential 1990

	Fossil fuels	Electricity	Cogeneration	Total
Conservation potential Mt	5.6	3.2	0.6	9.4
% of total conservation	18	13	—	19

Source: CEREN, 'Gisement des actions de maîtrise de l'énérgie dans l'industrie', ADEME, Paris, 1993.

Table 2 Conservation potential 1990 to 2005, Mtoe

Conservation potential	Economic growth impact	Added RD&D potential	Business-as-usual savings	Residual potential
9.4	+0.6	+1.5	–3.9	7.6

Source: CEREN, 'Gisement des actions de maîtrise de l'énérgie dans l'industrie', ADEME, Paris, 1993.

residual potential is the area on which a reinforced conservation policy, closely linked to the global warming issue, should focus.

The 1990 conservation potential of 8.8Mtoe (excluding cogeneration) is made up of 4.2Mtoe of fossil fuels and 1.9Mtoe of electricity in industrial processes, and of 1.4Mtoe of fossil fuels and 1.3Mtoe of electricity from other industrial uses. A greater actual amount of energy can be saved through actions directly related to production processes, so that savings in either electricity or fossil fuels can make up a larger proportion of energy consumption, totalling around 40 per cent.

Economics of industrial energy saving
The cost of conservation is FF3,500/toe. This means that an investment of FF3,500 in efficient technology will save one tonne of oil equivalent annually. Assuming an 8 per cent discount rate and a 10-year life time for replacing capital stock, this is equivalent to FF520/toe.

The total investment cost of implementing the conservation potential is estimated as FF29bn, with further FF10bn for cogeneration. Energy conservation costs for electricity are lower for measures not directly related to production processes, amounting to FF3,500/toe, whereas for potential process energy savings in electricity the average costs are FF4,300. As regards potential savings in fossil fuels, costs are very similar for both non-process and process activities: FF3,400/toe and FF3,300/toe respectively. See Table 3.[3]

According to the current energy prices in French industry – FF915/toe for fossil fuels (average price of coal, oil and gas, excluding tax) and FF1,415/toe for electricity – the average pay-back times for process energy conservation are 3.6 years (fossil fuels) and 3 years (electricity). For energy conservation on other uses, the pay-back times are 3.8 years (fossil fuels) and 2.5 years (electricity).

The range of conservation costs is wide, as can be observed from the cost curves. (See Figure 1.) The study looked at 77 conservation measures within process and 29 outside for which the investment costs for energy savings are available and are representative. The range of costs is summarized in Figure 2. Almost 40 per cent of the potential is available at less than FF3250/toe at current prices. One-quarter of the process potential is more expensive, costing around FF5,500/toe.

The criterion of pay-back time gives an initial view of the economic constraints which limit the implementation of energy conservation measures. See Figure 3. Industry tends not to invest in energy efficiency operations with more than two to three years pay-back time. When parameters other than energy itself – such as environmental or safety issues – play no part in the cost advantage, the acceptable pay-back time is often only two years or less.

The 23 subsectors for which the analysis was carried out were grouped into six sectors: the iron and steel industry; heavy chemicals; non-ferrous metals, lime and cement; metals (excluding metallurgy); consumption goods and materials (excluding cement); and the food industry. The aim was to create groups that brought together subsectors with similar criteria. These criteria were: relative homogeneity of cost of energy saving; the weight of potential energy savings in their energy consumption; and the foreseen trend

---

[3] For all uses (i.e. production process and other uses) the costs are FF3,300 for fossil fuels and FF3,900 for electricity, meaning FF3,500 for fossil fuels and electricity together, as in the table.

Table 3 Energy conservation potential in 1990 and related investment costs

	Process industries		Outside process industries		Total		Total
	Fossil fuels (boilers included)	Electricity (boilers excluded)	Fossil fuels	Electricity	Fossil (boilers included)	Electricity (boilers excluded)	
Energy consumption, ktoe	25155	18791	3544	2904	28699	21695	50314
Conservation potential, ktoe	4195	1894	1448	1300	5643	3194	8837
Potential/consumption	15%	8%	41%	45%	18%	13%	16%
Investment, mFF	13320*	6335*	4965	4555	18285*	10890	29175*
Conservation cost, FF/toe	3300*	4300*	3400	3500	3300*	3900*	3500*

* Recycling (glass and non-ferrous metals) excluded; (1kWh=2.22th,except for elexctric boilers 1kWh=1th).
Source: CEREN, 'Gisement des actions de maîtrise de l'énérgie dans l'industrie', ADEME, Paris, 1993.

Figure 1 Investment cost of conservation measures

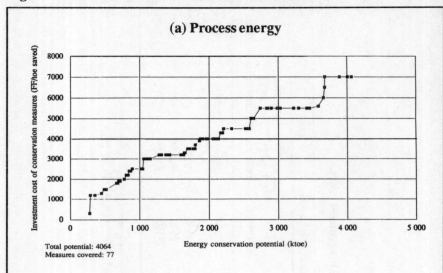

(a) Process energy

Total potential: 4064
Measures covered: 77

(b) Non-process energy

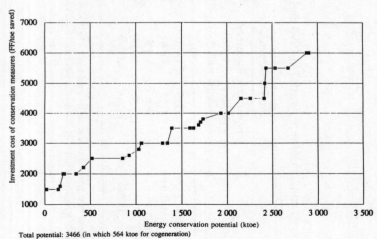

Total potential: 3466 (in which 564 ktoe for cogeneration)
Meaures covered: 34

Source: CEREN, 'Gisement des actions de maîtrise de l'énérgie dans l'industrie', ADEME, Paris, 1993.

Figure 2 Total industrial energy conservation

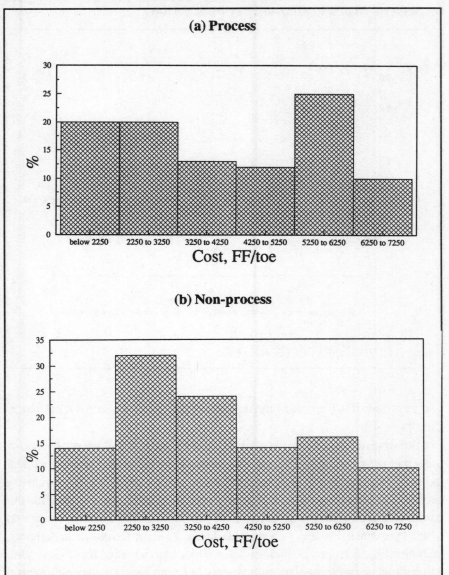

Source: CEREN, 'Gisement des actions de maîtrise de l'énérgie dans l'industrie', ADEME, Paris, 1993.

**Figure 3 Energy saving technologies in the French industry: correlation
between speed of adoption and pay-back period**

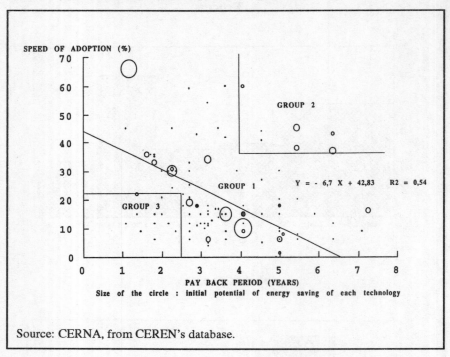

Source: CERNA, from CEREN's database.

of energy saved in the period and the kind of possible conservation measures.
(See Table 4.)

In the steel industry energy conservation costs are low and the prospects for
implementation are high. The opposite situation prevails in the heavy chemical
industry – high costs and low prospects for implementation. The food industry
is in a similar situation to the heavy chemical industry except that the
conservation potential is quite large: 28 per cent of consumption compared
with 12 per cent in the heavy chemical industry. The non-ferrous metal industry,
including lime and cement, is characterized by a high potential for conservation
and by higher implementation prospects but with two very special central
measures concerning the recycling of non-ferrous metals and waste incineration
in cement furnaces.

Table 4 Energy conservation potential by industrial sectors

	Energy consumption in 1990 ktoe	Conservation potential ktoe	Potential/ consumption	Conservation total cost mFF*	Conservation cost/toe FF/toe*	Prospects of implementation of energy conservation 1990–2005 ktoe	Residual potential ktoe
Steel industry	9656	1394	14%	3210	2300	1096	627
Nonferrous metals, cement	5756	1614	28%	3025	2900†	835	1161
Metals, metallurgy excluded	6868	1561	23%	4885	3300	465	1347
Materials including general consumption goods, but excluding cement	12253	2060	15%	6470	3500	744	2076
Food industry	5618	1573	28%	6385	4400	285	1516
Heavy chemicals	10243	1198	12%	5200	5100	492	913
Total industry	50394	9401	19%	29175	3500	3917	7640

* Cogeneration and recycling excluded.
† FF3,800/toe, waste recovery in cement industry excluded.
Source: CEREN, 'Gisement des actions de maîtrise de l'énérgie dans l'industrie', ADEME, Paris, 1993.

Three sectors have a bigger share in conservation potential than in energy consumption: they are the non-ferrous metal industry, the metal industry excluding metallurgy, and the food industry. The three others – the steel industry, consumption goods and other materials and heavy chemicals – have a larger share of consumption than of potential. In the steel industry and in the non-ferrous metals/cement industry, the proportion of conservation measures implemented are highest. These two sectors provide half of the energy saved from 1990 to 2005. In heavy chemicals the implementation rate averages, 35 per cent. The metal industry (excluding metallurgy) and industry producing consumption goods and materials implement only 26 per cent of their potential, under the study's assumptions. The rate is 16 per cent in the food industry. In such a situation, the food industry sector and the consumption goods and materials sector make up around half of the residual potential in 2005.

The CERNA survey: CO_2 emission taxation and competitiveness of French industry

As already mentioned, a second study commissioned by the French government was carried out by CERNA, ADEME and the French Ministry of the Environment. It focused on the perceived consequences for major energy-consuming industrial firms in France of the CO_2/energy tax proposed by the Commission.[4]

The approach of the CERNA survey is sectoral and microeconomic. The breakdown of data relating to energy consumption by the various branches of industry and on foreign trade allows accurate quantification. The analysis is micro economic, involving a survey of 22 firms, and does not enable any quantified aggregates to be established. However, it enables a typology to be created, which considers the differing attitudes among companies to the tax. The general trends to emerge from their reactions were pinpointed and the scale of their reactions was assessed.

A number of calculations were made concerning the methods of offering compensation or enabling firms to adapt to this tax, but this issue has not

---

[4] The tax referred to was presented in June 1992 as doc COM 92 226 published in the official journal of the EC on 3 August 1992.

been studied in depth. Likewise, the overall effects of such a tax on the material and energy content of economic growth have not been dealt with, nor has there been consideration of the repercussions of these effects for industries with high energy consumption – which are generally upstream of the production system – or for the industrial sector as a whole.

The effects on French industry of the tax proposed by the European Commission

- A tax rising gradually to $10/boe by the year 2000 is too low to result in any changes in the direction of the technological trajectories in the industries which are major energy consumers.

- Its direct impact upon CO_2 emissions within these industries would, therefore, be fairly restricted. Any accurate assessment of these effects was impossible within the framework of the study: it would have required an assessment not only of the effect of the tax on the extent and the rate at which the static potential for energy savings would be exploited, but also of the impact it would have on technical progress. Not only is any such assessment difficult in itself but it presupposes (erroneously) that the post-2000 tax scenario has been mapped out by the Commission.

- While having only a limited direct impact on curbing emissions, the tax would affect the competitive capacity of a number of industries operating in France. The branches hardest hit – those where the tax would exceed 2 per cent of value added – account for 24 per cent of industrial value added. Even if the tax were to be offset by lower social security charges, a number of sectors would still have to bear a levy representing over 5 per cent of their value added. This is the case for cements and lime, steel, fertilizers, aluminium, chemical electrolysis, fero-alloys, alcohol distillation, sugar, tiles and bricks, zinc and lead metallurgy, and the processing of cereals.

- Firms would react to the tax in different ways. Given the competitive environment in which they operate, some would be able to pass on the tax (in whole or in part) via the prices they charge, thus causing downstream substitutions, between materials or between products: these are among the positive effects expected from the tax. Some would lose market shares to firms established in France which produce competing goods with a lower energy content: this would be another of the anticipated positive consequences.

Sharp falls in activity to the benefit of competitors abroad (outside the tax area or within it but enjoying more favourable energy supply conditions or some technical advantage) would, in fact, occur only infrequently.

The tax would, however, hasten the trend towards geographical transfer which is an indication of how major French firms are becoming increasingly global. This trend would mean that each firm would concentrate its production activities at the most favourable sites. The planned dates for closure of facilities in France would be brought forward and there would be a dearth of new investment in productive capacity within the country. All these trends would be signs of France's loss of attractiveness as a result of the tax although firms themselves might not be seriously affected if they moved their heavily taxable activities abroad. Nevertheless this weakening of the industrial base would have to be offset by making France more attractive for other industrial activities. This could be a difficult undertaking, particularly as goods industries – the bulk of those most severely hit by the tax – form one of the leading sectors at present.

• Classic protectionist measures at the borders of the taxed area are unlikely to be established, since they appear to be impracticable both technically and politically. There is widespread doubt and uncertainty in industrial circles as to whether competitive differentials can indeed be redressed. The tax would, therefore, inevitably jeopardize the interests of the tax area whenever global firms were required to take decisions regarding the geographical location of new plants.

Are sector voluntary agreements an alternative to taxation?

The CERNA analysis suggests that sector voluntary agreements are an instrument for achieving reductions in CO_2 emissions whose effects would be very different from those produced by taxation. Voluntary agreements or contracts could be drawn up between an administration and a group of industrial firms. The agreements would aim to reduce CO_2 emissions by a certain amount by a specific point in time. The firms would be free to organize how they would share the reduction and what measures they would choose to reach their goals. Compared to the tax the contract approach has four main shortcomings :

• Contracts fail to ensure that the total direct cost of a set of actions undertaken to curb emissions would be minimalized, whereas a tax, relying on competitive decision-making, would ensure this.

• The amount of emissions necessary for the production of a particular item would not affect its price as it would if a tax were levied. Therefore, price-induced switching to 'cleaner' products and materials would not take place.

• The dynamic impact on technical progress would be slight.

Sustainable Energy Policies:
Political Engineering of a Long Lasting Consensus

Ernst Ulrich von Weizsäcker

Abstract

Energy industries think long term. Their 1993 investments should reflect what we believe to be the realities of 2020. Responsible managers even have the year 2040 in their sights.

Long-term prospects for the global climate appear threatening. The Intergovernmental Panel on Climate Change (IPCC) recommends 60 per cent reductions in global greenhouse gas (GHG) emissions within about 50 years from now. From a global equity standpoint this may imply a reduction of GHG emissions of the order of approximately 80 per cent for OECD countries. Nuclear energy and renewables cannot fill the gap. The only realistic way out is a dramatic increase in energy productivity (with renewables accounting for a growing share).

Given optimum conditions, macroeconomic energy productivity could increase fourfold between now and 2040. The task of engineering these conditions is principally a political one. Two – not mutually exclusive – options are available: the bureaucratic approach (through efficiency standards) and the market approach (through prices). I believe the market approach is both more powerful and more socially acceptable. The favoured instrument is a well-designed, gradual, long-term, revenue-neutral, ecological tax reform. Prices of fossil and nuclear energy can be raised sufficiently to induce technology, logistics and, indeed, our entire civilization to raise energy productivity by at least 3 per cent per annum over a period of at least 40 years. According to the World Resources Institute (WRI), ecological tax reform is bound to make our economies stronger, not weaker.

The strategy of pushing energy productivity is meant to serve as the basis for a long lasting energy consensus. It is likely to draw its supporters from all political camps, from all trades and from all countries including less developed countries which have found it difficult to adopt any stringent measures of (costly) pollution control. The EU has made a proposal for taxing energy and carbon dioxide emissions which represents a bold step forward, but still leaves considerable scope for improvement.

For the energy industry, the strategy involves a shift from selling energy to selling energy services – a growing trend in any case, as least-cost planning concepts are spreading throughout the world. The energy business in a wider sense could benefit commercially from a social strategy which stresses the importance of increasing energy productivity.

An alliance of long-termers

At international meetings, the representatives of the energy business always strike me as genuine advocates of long-term thinking. That's what I like about them. I simply assume that it should be possible to form an alliance with them, an alliance of long-termers. Other long-termers include educators, infrastructure planners, the wise men and women who exist in every society. The Club of Rome may also qualify as part of the alliance of long-termers, since all its noteworthy reports have placed a distinct emphasis on the long-term consequences of what the world is currently doing. The first report to the Club, the famous *Limits to Growth* of 1972, laid the foundations to what was later to be known as sustainable development.[1]

Sustainability, too, is a symbol of long-termism. Since the publication in 1987 of the Brundtland Report the international debate has definitely seen a shift from short-term crisis management to a more long-term perspective.[2] Sustainability has become the most important yardstick for desirable development, although yardstick is actually not quite the right term because nobody has so far managed to measure sustainability.

Nevertheless we can judge with some certainty processes that are not sustainable. Among them are the burning of fossil fuels at present rates, current rates of population growth, or the rate at which new states are being formed in eastern Europe.

Sustainable energy policies clearly require long-term thinking. Energy investments tend to require an extended planning period, a couple of years for construction and a long period of time for the use of the investment. Stable economic conditions, predictable demand, secured fuel supplies and social acceptability are the ingredients for commercial success in the energy business.

On the other side of the intended alliance, the environmental camp, the aim is long-term ecological stability. Following the success of pollution control – in the 1970s in the US and Japan and the 1980s in Germany and other European countries – which was not really a long-term affair, but rather one of crisis management, we environmentalists are turning our attention to such long-

---

[1] Donella H. Meadows (ed), *Limits to Growth: A Report for the Club of Rome's Project on the Predicament of Mankind*, Universe Books, New York, 1972 (second edition 1989).
[2] United Nations World Commission on Environment and Development, *Our Common Future*, United Nations, New York, 1987 (The Brundtland Report).

term issues as the greenhouse effect, long-term disposal of nuclear waste, genetic erosion,[3] the ozone layer, population growth and changes in our lifestyles.

To begin with the last of these, the lifestyles prevailing in OECD countries are very far from being sustainable. One thousand Germans consume roughly ten times more energy and other limited resources per capita than one thousand people in less developed countries. If German – or North American – consumption levels were the precondition for political stability, there would be no hope left for the global environment, because it is economically unfeasible and ecologically impossible to have such high consumption levels copied by 5.5 billion people. (See Figure 1.)

People around the world have begun to realize that we are on a dangerously unsustainable trip. I invite the energy community to agree with this statement and to join hands with us environmentalists in seeking realistic long-term strategies to bring us out of the present crisis.

The greenhouse effect

The greenhouse effect provides a prime example of the magnitude of the crises we are facing.

The Intergovernmental Panel on Climatic Change (IPCC) sees a reduction by some 60 to 80 per cent of greenhouse gas (GHG) emissions as a necessary target to prevent a dangerous acceleration of global warming. What the climatic experts may still have underestimated is the threat of sudden rises in sea level. The polar ice caps may be less stable than thus far assumed, and past rises in sea level seem to have occurred in rather dramatic jumps of a couple of metres within decades.[4]

Contrasting sharply with the needs established by the IPCC, some energy experts forecast increases in energy demand worldwide of the order of 50 to 70 per cent until 2020, with perhaps a doubling by 2040. Even the most

---

[3] Genetic erosion refers to loss of biodiversity, but specifically with regard to genes rather than habitat.

[4] Michael Tooley, 'The Impacts of Sea-Level Changes – Past and Future', in *Changing Weather Patterns*, Gerling Global Reinsurance Company, Cologne, 1990.

Figure 1 Comparative resource consumption

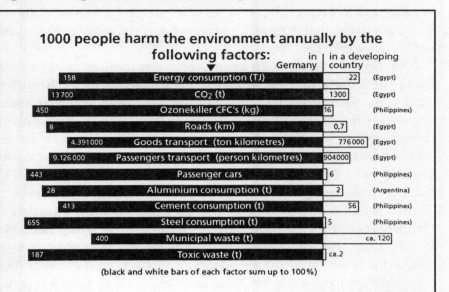

1000 people harm the environment annually by the following factors:

	in Germany	in a developing country
Energy consumption (TJ)	158	22 (Egypt)
CO_2 (t)	13 700	1300 (Egypt)
Ozonekiller CFC's (kg)	450	16 (Philippines)
Roads (km)	8	0,7 (Egypt)
Goods transport (ton kilometres)	4.391000	776 000 (Egypt)
Passengers transport (person kilometres)	9.126 000	904000 (Egypt)
Passenger cars	443	6 (Philippines)
Aluminium consumption (t)	28	2 (Argentina)
Cement consumption (t)	413	56 (Philippines)
Steel consumption (t)	655	5 (Philippines)
Municipal waste (t)	400	ca. 120
Toxic waste (t)	187	ca.2

(black and white bars of each factor sum up to 100%)

Source: R. Bleischwitz and H. Schütz, *Unser trügerischer Wohlstand. Ein Beitrag zu einer deutschen Ökobilanz*, Wuppertal Institute for Climate, Environment and Energy, Wuppertal, 1992, reproduced from Ernst Ulrich von Weizsäcker, *Earth Politics*, Zed Books, London, 1994.

aggressive build-up of nuclear power and renewable energy sources – at very high ecological and economic cost – would not be able to close the gap left between the IPCC exigency and GHG emissions associated with typical forecasts of energy demand.

No classical political measures are available to respond to this extraordinary challenge. Desirable and necessary measures – among them control of CO_2 emission standards; forest protection; subsidies for afforestation, dry rice cultivation (to reduce methane emissions from rice paddies), or solar technologies; behavioural education; and ecological management – are still bound to miss the target by a wide margin. The climate challenge requires new thinking on a very fundamental level.

A new direction for technological progress

We won't get a cheaper solution to the problem than by fundamentally re-directing technological progress. This means that values and cultural progress will also have to be redirected.

Technological progress in the past has been characterized chiefly by the increase of labour productivity. Science and technology together with logistics and good management have enabled the pioneering countries to increase labour productivity roughly by a factor of 20 over a period of 150 years. Resource productivity was lost out in the process, increasing hardly at all, as can be seen from the fact that energy and material resource consumption grew almost in parallel with GDP in all industrialized countries. It was only after 1973 that a moderate decoupling occurred, triggered by the oil price increases.

Now it is time to change the course, in the direction of 'eco-efficiency'.[5] The inventors of this term – Schmidheiny's Business Council for Sustainable Development – did not undertake to quantify the efficiency objectives. In terms of macroeconomic energy productivity, we must aim at a factor of four to close the gap identified above. What the world needs we at the Wuppertal Institute have called the efficiency revolution.

Quadrupling energy productivity – that 'revolutionary' objective – is not as outlandish as it may at first sound. It could be reached by a mere 3 per cent annual increase over some 45 years. And for many processes involving energy consumption a doubling of efficiency is possible even using existing tech-nologies and without requiring any major changes in behaviour or infra-structure. If more extensive changes of this sort are allowed, a quadrupling of macroeconomic energy productivity is conceivable with the use of existing technologies.

The point of entry into rising energy productivity will be least cost planning (LCP). If power plant permits are made dependent on the proof that there are no lower-cost alternatives available to fill the expected energy gap, it will soon become more profitable for the utilities to subsidize energy efficiency at the point of consumption. Utilities are able to raise the price per kilowatthour

---

[5] Stephen Schmidheiny and the Business Council for Sustainable Development, *Changing Course: A Global Business Perspective on Development and the Environment*, MIT Press, Cambridge, MA, 1992.

if the monthly bills of their customers are reduced. The profits from such efficiency investments will thus be shared between utilities and their customers. The capital savings for the utilities can be very substantial. Pacific Gas and Electric, one of the largest utility companies in the world, has closed its construction department and increased its profits. Similar developments are under way at Ontario Hydro, Canada's largest energy company.

However exciting this development is, it will not reach beyond relatively superficial savings to reduce excessive waste of energy. Truly new energy-efficient technologies which involve new philosophies of low-energy intensive food production, durability of goods and new infrastructures that help to reduce the energy bill are not to be expected as a result of local LCP, simply because energy prices are too low to make major changes profitable.

Ecological tax reform

The straightforward way of driving technology development into a new direction is to make it profitable. Energy prices should rise. Energy efficiency had its heyday during the 'energy crisis' in the early 1970s. Prices can be raised in any of four ways: by bureaucratic measures; by cutting subsidies; by establishing a regime of tradeable energy permits; or by raising taxes.

Tradeable permits should be the preferred option for *international* negotiations.[6] A system of emissions permits based on assumptions of equity per capita can theoretically be established. The south would clearly benefit from such an allocation and would try to gain as much revenue as possible from selling permits to the north. Some concessionary 'grandfathering' could be negotiated to reduce costs for the north,[7] however, I am rather sceptical as to whether the north would really accept a regime that turns out to be *very* costly for northern economies.

Nevertheless, it is fairly difficult to imagine how a system of tradeable permits could work in *domestic* markets. Who would measure and control carbon dioxide emissions from wood stoves or methane emissions from compost?

---

[6] Michael Grubb, *The Greenhouse Effect: Negotiating Targets*, RIIA, London, 1989 (second edition 1992).
[7] 'Grandfathering' assumes that we have inherited a 'right' to pollute; permits are allocated in proportion to current or previous emissions.

The establishment of a carbon dioxide market even for fossil fuels would involve considerable additional monitoring and control costs. Moreover, permit prices would be likely to increase sharply in economic boom times, causing severe social problems for the poor who might enjoy no benefits from the economic upswing, so that they – or their suppliers – might find themselves unable to pay the high prices for permits.

By comparison, problems such as administrative cost, unpredictability of prices and social equity can be greatly reduced by a strategy of direct pricing of carbon dioxide emissions or of energy. I shall now outline such a strategy, which may be labelled 'ecological tax reform'.

The strategy would begin by reducing subsidies on basic commodities. Why should coal mining, nuclear energy or agricultural products be subsidized by taxpayers' money? Even 'subsidies' (i.e. public spending) for roads or airports are ultimately irrational. He who wants transport ought to pay the full price. It is an economic fallacy to believe that transport *per se* is a good thing for the economy as a whole. It benefits only volume and turnover of the economy. But, then, even traffic accidents add to 'economic' turnover without making anybody happier.

Cutting subsidies is not enough. Externalities should also be accounted for and incorporated into prices. To this end, taxes may be levied on non-renewable sources of energy, on primary raw materials, on water consumption, on certain chemicals such as chlorine or metals, or on certain types of land use. (I remain sceptical of taxing polluting emissions or waste, because of the problems of evasion and control which would arise, chiefly in less developed countries.)

Externalities need not be calculated in detail. Crude estimates suffice. Taxes have the advantage of not requiring erudite justifications via the computation of externalities. Income taxes or VAT were never justified in that way. But, according to Pigou, the economy as a whole would benefit if taxes were to more or less reflect the costs to society of energy and other basic commodities.[8] Studies such as the one on energy externalities commissioned by the German Ministry of Economic Affairs seem to indicate that externalities amount to approximately 10 per cent of GDP.[9]

---

[8] Arthur Cecil Pigou, *The Economics of Welfare*, Macmillan, London, 1920.
[9] Prognos, *Externkosten der Energieerzeugung*, Study (6 Vols) for the Federal Economic Ministry, Bonn, 1992 (in German).

Figure 2 Profitability shift resulting from ecological tax reform

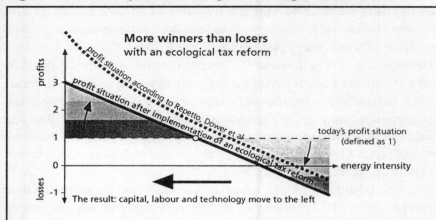

Notes: Current levels of profit are defined as unity over an axis of energy or pollution intensity. Gradual introduction of environmental tax reform will lead to a gradual decline into the red for bulky, energy intensive or polluting operations, while profits for relatively clean activities will improve. As a result, investment capital and technology and labour will move from right to left.

Source: Ernst Ulrich von Weizsäcker, *Earth Politics*, Zed Books, London, 1994.

To avoid any expansion of public spending, other taxes, charges and levies should be reduced by equivalent amounts. There is a particular need to reduce the fiscal and parafiscal burdens on human labour, thus making labour more affordable again for employers. (See Figure 2.)

The cutting of subsidies together with ecological tax reform would gradually make the increase of resource productivity more profitable and the dismissal of workers less so. Repetto and his co-authors, quoting Ballard and Medema, seem to have proved Pigou's assumption and state that:[10] 'the total possible gain from shifting to environmental charges could easily be $0.45 to $0.80

---

[10] Robert Repetto, Roger C. Dower, Robin Jenkins and Jacqueline Geoghean, *Green Fees: How a Tax Shift Can Work for the Environment and the Economy*, World Resources Institute, Washington DC, 1992; Charles L. Ballard and Steven G. Medema, 'The Marginal Efficiency Effects of Taxes and Subsidies in the Presence of Externalities: A Computational General Equilibrium Approach', Department of Economics, Michigan State University, East Lansing, MI, 1992.

per dollar of tax shifted from "goods" to "bads" – with no loss of revenues.'
Actually there seems to be empirical evidence for this result, namely that
countries maintaining high energy prices have fared better in economic terms
than those with low energy prices. (See Figure 3.)

The greening of subsidies and ecological tax reform may be designed in
such a way that end user prices for ecologically problematic factors increase
steadily and predictably by constant factors over many decades.[11] A reasonable
first approximation may be a price increase by 5 per cent annually which
would lead to a doubling in 14 years, a quadrupling in 28 years and an eight-
fold increase in 42 years. This is a very strong signal which can easily change
the course of technological progress.

And yet, owing in part to expected productivity gains (estimated
conservatively to total 3 per cent annually), the price signal would be extremely
weak, amounting to 2 per cent (5 per cent minus 3 per cent) annually for a
production factor weighing on average less than 4 per cent of total production
costs, so that the remaining cost differential would be only 0.08 per cent
annually.[12] Even this almost imperceptible cost differential would – on average
– be outstripped by labour cost reductions (for instance, from reduced social
security payments). Hence, firms would on average actually benefit financially
from the reform.

To be sure, there would be losers. Aluminium smelters (from bauxite), steel,
bulk chemicals, cement, pulp and paper and a few other industries would run
into difficulties even if the price signal remained as weak as suggested. To
avoid additional unemployment, consideration would have to be given to
allowing such firms a temporary tax relief until their existing capital stock
was written off. But investments into the same old-fashioned production should
be prevented.

If the World Resources Institute (WRI) researchers are right, one should
expect far more winners than losers. I am convinced that energy companies
can restructure with comparative ease in a way that makes them winners as

---

[11] Ernst Ulrich von Weizsäcker and Jochen Jesinghaus, *Ecological Tax Reform: Policy
Proposal for Sustainable Development*, Zed Books, London, 1992. For a broader outlook:
E.U. von Weizsäcker, *Earth Politics*, Zed Books, London, 1994.
[12] It is assumed that efficiency gains would be larger in eastern Europe and comparable in
developing countries.

Figure 3 Economic performance and energy prices 1978–90

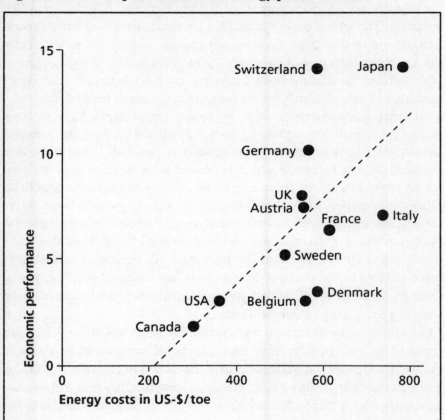

Notes: Economic performance being defined by a composite parameter consisting of rank scores for trade surplus (or deficit), patents per one million inhabitants, GDP growth and energy efficiency. Energy prices are defined by a mix of electricity for industry, petrol and light oil for domestic heating. There seems to be rather a positive than a negative correlation between economic performance and energy prices.

Source: Rudolf Reichsteiner, 'Sind hohe Energiepreise volkswirtschaftlich ungesund?', *GAIA*, Vol. 2, 1993, pp. 310–27, reproduced from Ernst Ulrich von Weizsäcker, *Earth Politics*, Zed Books, London, 1994.

well. The centrepiece of their future business would be energy services: selling packages of energy management for homes, factories, hospitals or office buildings. The present energy companies are among the most skilled experts identifiable in this field. Gasoline companies (which are now already diversifying widely) could move into the mobility market, competing there with logistics firms, car manufacturers and car rental services. They could specialize in fuel efficiency for the car and lorry fleets of larger firms.

In the political arena, the tax strategy presents two main problems: whether elected governments can reliably implement a long- term strategy, and whether international harmonization of the strategy is possible. But if there is a consensus in society that the scheme is beneficial for the economy – not only for the environment – its acceptance by all major political parties should be fairly easily gained, so that it would never become an election issue. International harmonization will have an easier passage than classical environmental policy measures, which invariably involve extra costs with no immediate economic benefits. There is also the point that the cutting of subsidies is recommended to less developed countries in any case, and simple resource taxes are far easier to collect than personal taxes as everybody who has ever lived in a developing country will admit.

In addition to the theoretical macroeconomic gains which the WRI team expects from ecological tax reform there may be still more attractive prospects involved in the idea.[13] If business leaders and owners of capital see scope for a new and reliable way towards technological progress they may feel encouraged to channel money in that direction. Few factors are more stimulating for the business world than a steady and predictable trend. Once the consensus is strong enough to persuade the pioneers, the process can soon become self-enforcing and may lead to a new Kondratieff cycle.[14]

Is it not an extremely tempting idea to use the high certainty of the environmental crisis as a basis for a high certainty of a new technological trend? Is it not, therefore, a good idea to put the shift of technological progress on the agendas of the G7, of the OECD or of the relevant UN bodies?

---

[13] Repetto et al., op. cit.
[14] Kondratieff cycles (named after their inventor) refer to long waves of some 50 years. 'A new Kondratieff cycle' denotes the (much desired) lasting upswing, not the downswing phase of the cycle.

Transport costs and international trade

Implementing the strategy of 'making prices tell the ecological truth' could have an important side effect. High energy and resource prices also mean higher transport prices. The transport sector has been a favourite target for state subsidies. As long as labour productivity increases were the main objective of technological progress, free use of infrastructure and low taxation on transport seemed quite rational. But once resource productivity becomes a parallel objective, excessive transport will be seen to be one of the main sources of wasteful resource use. (Figure 4 shows how transport fuel usage correlates to fuel prices.)

International trade benefits greatly from the infrastructures once funded by taxpayers. Development aid has for a long time favoured the construction of ports, roads, railways and airports. Primary commodities thus became much more accessible, and international competition has contributed to an increase in commodity supplies and gradually to lower prices.

It is now the time to recognize that subsidized transport costs can actually be destructive not only to the environment but also to the economy.[15] What is the use of spending scarce public funds to make foreign goods artificially cheaper than they would otherwise be?[16] And what is the point of developing countries specializing, to an ever-increasing extent, in commodity exports which contribute very little to development? Higher transport prices will render unprofitable the criss-crossing of the oceans by low-value goods and will direct international trade towards high-value goods.

Making transport prices tell the ecological truth is perhaps the only legitimate 'protectionism' and the one that is also least bureaucratic and discriminatory.

A new equilibrium will have to emerge between economies of scale and resource efficiency. The amount of manufacturing that is contracted out will decrease again as the supply of parts from very distant places becomes an apparent absurdity. (It's a macroeconomic absurdity *now* in many cases, but appears rational on the microeconomic levels.)

---

[15] Currently fuel for ships and planes is not taxed. Ports and airports tend to be funded by public money, often development aid.

[16] Travel would also be more expensive for non-commercial users and domestic traders, but they would only be paying the full costs of their actions rather than letting the taxpayer pay as currently happens.

Figure 4 Annual per capita motor fuel consumption plotted against fuel prices in OECD countries 1988

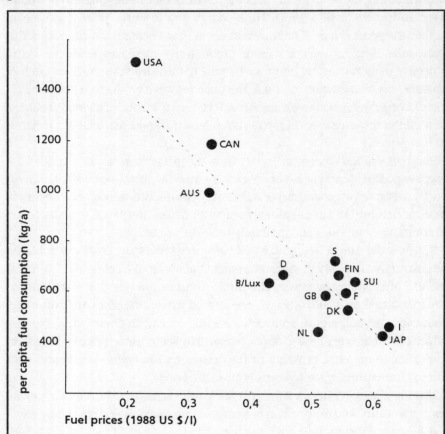

Notes: The graph seems to indicate a very strong, if long-term, price elasticity for the use of motor fuels.

Source: Ernst Ulrich von Weizsäcker and Jochen Jesinghaus, *Ecological Tax Reform: A Policy Proposal for Sustainable Development*, a study prepared for Stephan Schmidheiny, Zed Books, London and Atlantic Highlands NJ, 1992, reproduced from Ernst Ulrich von Weizsäcker, *Earth Politics*, Zed Books, London, 1994.

Rising transport costs may give some relief to industries and farms suffering ruinous competition from abroad. Local markets are likely to regain some economic importance which may also be good for social cohesion.

For primary commodity exporting countries including OPEC and for certain industrial branches such as bulk chemicals, cement, metals and long-distance haulage, the shift may at first look unpleasant. However they cannot count on maintaining the present development for ever, and a slow and well-planned phase-out of unsustainable processes may be the best they can get. Following the example of the North American utilities, they may even (with their specific knowledge) become active – and profit-making – players in the transformation process. They may become the first suppliers of primary commodity-related services and make increased profits through using less of those commodities.

At the international level, some compensation measures for primary commodity exporting countries are quite conceivable. As I am assuming that the northern economies would benefit greatly from a strategy of shifting towards resource productivity I see no reason why the north cannot afford some additional official development aid directed to the necessary restructuring of former monostructures of primary commodity exporters.

The Restructuring of the Electric Utility: Technology Forces, R&D and Sustainability

Carl J. Weinberg

Abstract

Major forces are restructuring the electric utility industry. The vertically integrated structure – generation, transmission, and distribution – is under assault and a major paradigm shift is under way. The change is driven principally by competition in electricity supply, customer energy efficiency, and environmental values. These forces are working together to create major changes, and behind each of the forces, a technology shift is under-way. The electricity supply industry is moving from boiler-based to turbine-based power plants. The energy efficiency of building designs, appliances, construction materials, lighting, air handling equipment and is increasing in every case, and reducing the projections of the rate of growth of electrical demand. Environmental concerns have continued to foster the development of renewable technologies and have introduced new values to be considered in future investments. Renewable technologies, particularly wind, have made great strides in the last decade. Cost projections and actual bids are bringing them into the circle of competitiveness with fossil resources. None of these technologies have reached maturity, and all are still on a path of rapid change.

The inherent investment basis for electric utilities is also changing since these new technologies follow a different economy of scale. They are all products and therefore follow economies of production and not economies of construction. The shift is from constructed energy to manufactured energy.

This will have a profound effect on planning, management, operation, capital efficiency and risks. There is a need for a new optimizing function that moves from the dominance of busbar energy cost to cost of service. This could change the electric utility from being a supplier of electrons to being a supplier of services, and, coupled with potential restructuring, will bring about a review of revenue mechanisms.

There is evidence that some of these forces are already at work. Gas turbines are dominating the independent power supply market; village power systems are being installed in place of grid extension; energy efficiency programmes are being driven globally by non-governmental organizations (NGOs); power generation is being separated from transmission and distribution. Research implications also exist. Generating systems will require market stimulation and market application to reach appropriate designs and cost targets. Products are also subject to more rapid technical change. Systems will have to be integrated all the way down to the distribution system. Energy efficiency may well be dispatched from customer linked gateways. Resource expansion will come to resemble portfolio investments. A portfolio analysis for a summer peaking utility indicates that using 30 per cent intermittent sources (such as wind) can result in lower costs than those incurred in the use of typical new generating technology, and in costs 5 per cent higher than those incurred in the use of advanced coal and gas technology, but with significantly lower emissions of carbon dioxide.

The route to sustainability will then be the resource investment portfolio, consisting of fossil fuels, energy efficiency and renewables, in a ratio depending upon local resource availability and local policies. As long as the good of society is judged important and the electrical energy market is not driven by short-term market forces, a portfolio approach will allow a gradual shift towards a more sustainable energy future.

A new value system

Sustainable development is a key facet of a new value system emerging both in the United States and around the world, many of whose central ideas were first defined for the global community in *Our Common Future,* a report issued in 1987 by the United Nations World Commission on Environment and Development.[1] The report argued that basic questions on the environment and economics can no longer be treated separately. It noted that economic policies that have assumed the existence of an unlimited and self-restoring biosphere must now change to recognize severe ecological limits. It recommended a strategy of 'sustainable development in which the material needs of all the world's people are met in ways that preserve the biosphere'. Sustainability, as a concept and as an objective, has important implications for utilities and for society. This paper reviews some of the dominant trends in the move towards sustainability and their implications for the future electric utility industry. Even though this paper focuses on investor-owned utilities as they exist in the United States, its content applies to other kinds of utilities as well, since all require significant capital investment.

One of the key changes in a worldwide value system is the shift from exploitation and consumption of resources to objectives of 'enoughness' and sustainability. 'Enoughness' means meeting a need or getting the job done in the manner that uses the fewest resources and has the fewest environmental impacts. It involves considering all options to meet a need and adopting the option that is just enough. It also entails giving consideration to full cycle, or 'cradle-to-grave' consequences of decisions and actions. People are talking more and more in terms of our obligation to be stewards of the environment and resources. This may well lead to a new definition of 'economic choice'. In the utility business, for example, a key sign of this shift is the widespread adoption of energy efficiency and energy conservation objectives and programmes. These programmes have economic and environmental objectives and advocate 'enoughness' in customer energy use and utility generation.

---

[1] United Nations World Commission on Environment and Development, *Our Common Future*, United Nations, New York, 1987 (also known as 'The Brundtland Report').

The traditional utility

The traditional utility is structured around large, remote, central station power plants that use transmission systems to deliver electricity to customers on the distribution system. (See Figure 1.) From the 1900s to the 1960s, power plants were built in increasing sizes to capture the economies of scale. The chosen strategy was to grow and build because each new plant reduced the busbar cost of electricity.[2] During the first hundred years of the development of the utility industry, tremendous economies of scale could be achieved by building ever larger central generation units. With significant gains to be obtained by investing in generation technology, it is not surprising that most utility research and modelling has concentrated on the central station generation approach. Since the 1960s, however, it has happened both that thermal efficiencies have flattened and economies of scale of large scale generation have been exhausted. Prior to the Second World War, power plants converted only 21.8 per cent of a fuel's energy content into electricity. With the assistance of advances in metallurgical knowledge and materials gained from development of new aircraft and artilleries during the Second World War, thermal efficiency rose to 32.9 per cent by 1965, and has since levelled off. Metallurgical weakness at high temperatures and pressures, unreliability of large plants, complexity of plant piping, and environmental concerns constitute today's new barriers to greater thermal efficiency growth. In contrast, recent work in electricity production using turbine technology based on jet aircraft engines (so-called aeroderivative turbines) shows signs of being able to provide high efficiencies even at modest sizes.

Economies of scale are desirable because they reduce the per kWh cost of supplying energy to customers, but they contribute to what is called the 'lumpy' investment problem. This refers to the inability of utilities perfectly to match yearly plant capacity additions of the size necessary to keep pace with load growth, but yet take advantage of economies of scale. Lumpy investment is a more (less) severe management problem when load growth is slower (faster).[3]

---

[2] Busbar electricity cost is the cost of electricity as delivered at the power plant.
[3] S. Peck, S. Chapel and S. Vejtasa, 'Evolving Technologies, Utility Incentives and Alternative Financing and Cost Recovery Methods', *Resources and Energy,* Vol. 7, 1985, pp. 1-12.

Figure 1 Traditional utility structure and US electricity production 1990

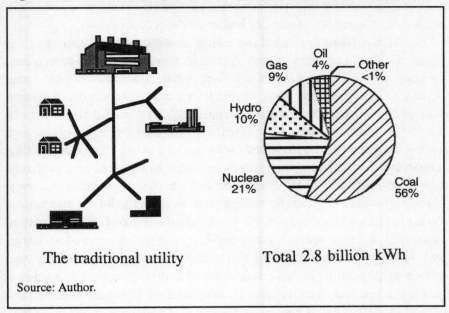

The traditional utility   Total 2.8 billion kWh

Source: Author.

As a result of exploiting economies of scale, most of the larger utilities now operate vertically integrated generation, transmission, and distribution systems. This feature, coupled with the monopoly franchise, was necessary to protect the long-term investment in these large facilities. The 'price' for protection was full regulation, national ownership, or some governmental supervision coupled with an obligation to serve – this obligation taking various forms either in regulation, government decrees, policy, or additional government funds to allow a utility to continue to operate. Electricity is a vital ingredient for social and economic development so that government really cannot allow a utility to become bankrupt or discontinue operation except in very unusual circumstances.

Utility technology runs its course

As long as the cost of electricity was decreasing, customers were satisfied, and so they were indifferent to the utilities' business practices. The customer bought kilowatt-hours with little concern about unit price and power quality. Likewise, the utilities had very little reason to learn how customers used their product. Consequently, electricity was treated as a commodity.

The 1970s marked the end of this utility model for doing business. (See Figure 2) The oil embargo triggered change. Reversing a long-term trend, the construction of bigger plants now increased busbar costs, although some of this cost came from growing environmental restrictions. Thermal efficiency increased from a few per cent in 1900 to around 40 per cent in the 1960s. Material limits have restricted increases since the 1960s, and additional gains will be made only with new materials or new competing technologies. Uncertainty in treatment of large, new investments probably also contributed to this peak in unit size. Utilities have been denied recovery of capital because the need for new generation capacity did not materialize during the time of construction. (While this analysis is directed primarily at fossil fuel plants, it seems equally applicable to hydro generation. Large hydro plant projects are becoming almost impossible to site and fund because of their widespread ecological consequences.) In the increasingly competitive global marketplace, the customer in industrialized countries became concerned about the cost of energy, and began using appliances and processes that were not only more energy-efficient but also more sensitive to power quality and reliability.

The stage was set for change. The period from the late 1970s to the end of the 1980s was a time for initiating changes in the utility industry. Those changes are still under way today.

Utility industry transformation

After almost a century of relative stability, the utility industry is changing. Forces, global and local, in regulation, technology, and markets are causing a fundamental restructuring of the gas and electric utilities. Hundreds of US utilities, some large – greater than 20,000 megawatts of electricity output (MWe) – some very small, some privately owned, and some public, are being forced to make adjustments. National utilities have been privatized, independents have been allowed to build power plants. The United Kingdom has privatized its electric utility industry; Sweden initiated privatization in

Figure 2 The typical fossil plant (using a boiler) has reached the top of its technology development S-curve

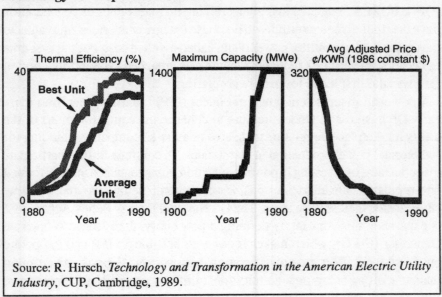

Source: R. Hirsch, *Technology and Transformation in the American Electric Utility Industry*, CUP, Cambridge, 1989.

1992; and Italy and Brazil are now privatizing. India is actively soliciting private power production. This trend will be global.

The forces driving these changes – competition, energy efficiency, and environmental sustainability – are certainly not unique to the electric utility industry. They are contributing factors to the increased emphasis on sustainability across commercial sectors and throughout society. They bring together the industry challenges (competition, growing environmental restrictions, rising costs associated with old technologies) and opportunities (new technologies and markets, and industry restructuring).

Competition in electricity generation

The Arab oil embargo of the 1970s disrupted many countries' fossil fuel supplies, raised energy prices, and contributed to an uncertainty about the energy future of the United States. In response to this, the US Congress passed the Public Utilities Regulatory Policy Act (PURPA) of 1978, which legislated for a new class of non-utility generating companies called qualifying facilities (QFs). PURPA mandated that plants relying on renewables or cogeneration have the right to require utilities to purchase the power at prices equivalent to the costs of the next utility power plant, called avoided cost. For the first time in decades, the utility's monopoly status was breached. The introduction of QFs was the first move towards restructuring.

A powerful independent power producer (IPP) industry grew from these early QF businesses. Under pressure to enhance the competitiveness of the utility marketplace, regulators mandated or allowed competitive bidding for 'wholesale' power production in many states. To continue fully to participate in the business of investing in power production, some regulated utilities formed non-regulated subsidiaries or businesses to compete outside and, in some cases, inside their service areas. The US National Energy Policy Act of 1992 is expanding this concept by defining a new entity, the Electric Wholesale Generator (EWG), which makes it easier to become an IPP and to operate outside the US. IPPs are now operating on a worldwide basis and defining their market as being outside traditional utilities and national boundaries. Who provides for new electricity generation will be a subject that each nation will have to decide, and will initiate a scrutiny of the industry that will have major impacts, and will bring all of the forces of change into consideration.

Now that the IPP industry is a substantial factor in building new generation capacity, the pressures are growing to open access to transmission. In some areas, open access is already available for wholesale purchases or transportation. US federal regulatory proposals call for a nationwide US access policy. US utilities must respond to any request for transmission access within 60 days. Some utilities are supporting and shaping the development of open access transmission regulation, while others are resisting it. The final operating rules for transmission are not yet settled. On a worldwide basis utilities have been reluctant to open access to transmission systems because of technical concerns. In no case were they successful.

In the United States, the gas utilities are already deregulated and no longer vertically integrated. Retail wheeling of natural gas is now a fact; that is, large customers can contract directly with gas suppliers, and the utilities must use their pipelines to deliver the gas. Electric utilities are concerned that retail transportation of electricity will be mandated in addition to wholesale wheeling. Because of the capital intensive nature of the electric utility industry, many utilities believe that retail wheeling will burden the smaller customers with undue fixed charges, or it will prevent investments from being recovered. They say the distribution system remains a natural monopoly. The US Congress so far has agreed with the utilities. Nonetheless, to one degree or another, the US electric utility industry appears to be heading for the dismantling of vertical integration. In many ways this mirrors what is happening in the UK, Norway and Sweden.

Gas turbine technology

High-efficiency gas turbines derived from aircraft jet engines have the potential to make improvements in generation efficiency. These advanced gas conversion technologies could achieve efficiencies in the range of 60 per cent (lower heating value). (See Figure 3.) The most pessimistic efficiency projections for these turbines are still higher than the operating efficiencies of the best of current gas-fired power plants. These turbines have the potential to reduce emissions, cut the cost of delivered energy, and lessen the impact of fuel price increases. Gas turbines are capturing a major share of the emerging independent power market. Combined cycle turbine power plants are rapidly becoming the system of choice. Demonstration and evaluation of even more efficient units are planned by the end of the 1990s.

Turbines are also important because they provide the means for efficient utilization of biomass. The first prototype, a six-megawatt biomass gasification combined cycle gas turbine, was recently commissioned in Värnamo, Sweden. A large 25MW plant is planned for Brazil's northeast state of Bahia, with the support of the Global Environment Facility (GEF).

Research is currently being undertaken to gasify solid fuels for use in gas turbines. This not only allows for increased system efficiencies but also provides for simpler emissions control techniques.

Figure 3 Conversion efficiencies for gas turbines and fuel cells

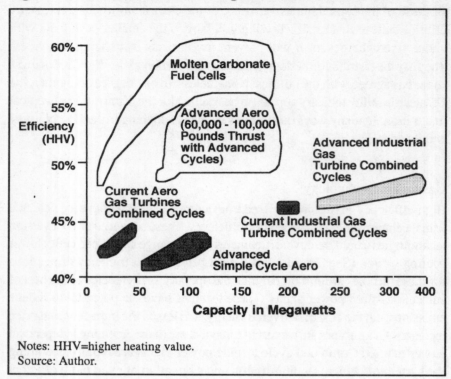

Notes: HHV=higher heating value.
Source: Author.

Customer energy efficiency

Concerns about dependence on imported oil, rising energy costs, and environmental degradation are prompting actions for increased energy efficiency. While the potential energy and economic savings are large – possibly as high as 75 per cent, according to some advocates – utility customers have adopted few measures beyond those required by US building standards. In the US, energy costs are still a relatively small fraction of most customers' budgets.

Many proponents of energy efficiency believe that the utility can be a highly effective vehicle for improving energy efficiency. Utilities have well-established and effective means of communicating with customers and can aggregate diffuse markets for energy efficiency equipment. State regulators are mandating that utilities manage customer energy efficiency programmes. In some cases,

Figure 4 US state regulators that have adopted or are considering incentives for utility programmes for customer efficiency

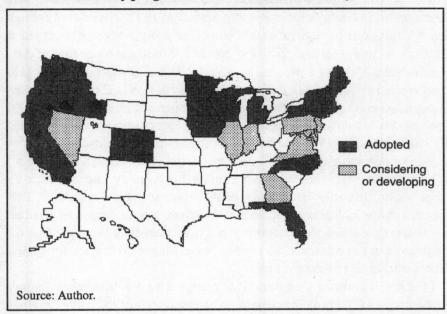

Adopted

Considering or developing

Source: Author.

regulators are also providing incentives to utilities to implement energy efficiency programmes for their own operations. More than half of the state regulators in the US have adopted, or are considering, energy efficiency incentives. (See Figure 4 and Sioshansi's paper above, 'The Pros and Cons of Learning by Doing'.) Energy efficiency is being advocated in many countries by non-governmental organizations to force governments to consider their programmes.

Energy efficiency poses a dilemma for utilities. If they are paid by the electron delivered, why should they want to deliver less electricity and decrease their revenues? Utilities, therefore, must be able to recover their expenditures by means of a higher cost of electricity. This clearly is economically efficient if the saving of electricity is less expensive than the production of electricity. But it causes a major shift from rates (costs per kWh) to bills (cost per kWh xkWh = bill). This is a fundamental change for utilities, and will require a different approach to capital recovery. A vigorous approach to efficiency, though, can also reduce the need to build generation facilities.

Energy efficiency measures

More than 600,000 of Pacific Gas and Electric's (PG&E) residential customers participated in energy efficiency programmes in 1992. They saved enough energy to power more than 90,000 homes annually, with yearly customer savings at current rates expected to exceed $75 million. More than 25,000 commercial, 10,000 agricultural, and 1,200 industrial customers also participated in various PG&E programmes. Together, PG&E customers saved 578 million kilowatt-hours of electricity and enough natural gas to serve 56,000 households annually. These participation and energy savings rates exceeded PG&E's goals for 1992 and bode well for achieving the overall target of reducing demand growth by 75 per cent by the year 2000. (See Figure 5.) These programmes, however, are coming increasingly under regulatory supervision, primarily focused on specific measurement and evaluation. This poses a philosophical problem, in that it will never be as easy or as accurate to measure the absence of electrons as it is to measure their presence. But, for regulators or governments, the need to prove what was obtained for capital spent is inherent in their politics.

PG&E's Advanced Customer Technology Test for Maximum Energy Efficiency (ACT^2) project is making real-world tests of the energy savings from a variety of energy efficiency measures. Data are collected on energy savings, technical performance, indoor environmental conditions, customer acceptance, and cost. Results to date indicate that significant energy savings – as much as 75 per cent of energy consumption – are technically achievable at costs competitive with energy supply, if appropriately designed packages are used.

It is important to emphasize the basic assumptions underlying the '75 per cent energy savings' statement. The first is the ability to spend funds equivalent to the present value of energy supply costs over time. This expenditure is not easy to calculate since it involves estimating supply costs over time and also estimating the potential life of the energy efficiency measures. The second is the use of appropriately designed packages. This entails combining a number of individual packages so that together they reduce energy savings. As an example, better fitting windows and improved insulation combine to reduce usage of air-conditioning or heating systems. This interaction is not well understood by either the architectural or engineering communities. Transaction

Figure 5 PG&E plans to meet 75 per cent of new electric demand by the year 2000, and a large fraction thereafter, with customer energy efficiency

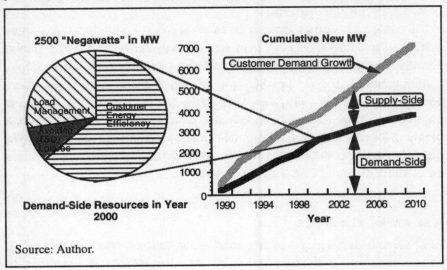

Source: Author.

costs of installation of energy efficiency measures can be large, to the extent that installation of only one such measure may be possible. Such measures need to be coordinated, however, and their interaction must be carefully considered and designed for.

The utility may well be the preferred infrastructure to take on the task of aggregating, coordinating and managing energy efficiency programmes. Restructuring of the utility industry will require careful consideration of the role of utilities in delivering energy efficiency. Only by properly defining that role can the technical potential be actually achieved.

Environmental sustainability

Most electricity generation causes some environmental damage. Dirty air, poisoned soil or water, land use impacts and displaced wildlife are some of the more typical consequences of fossil fuel, nuclear and hydro plants. At first, the costs, called 'externalities', were paid by the victims rather than by those who produced and used the product. Environmental movements since the 1960s have forced many of the generators to 'clean up their act'. The resulting increase in the cost of electricity meant that some of the environmental costs were being internalized.

While these costs offset some of the external costs of pollution, many argue that they do not fully reflect the environmental impacts of electricity generation and other business operations in market decisions. (See Figure 6.) For example, some quantities of NO_x and SO_x are still being released, as is the more controversial CO_2. In 1989, New York state attempted to 'internalize' these costs by requiring values to be added to the projected cost of generation when planning for new resources. These add-ons ranged from 0.25 to 0.005¢/kWh, depending on the pollutant.[4] For coal, probably the dirtiest source of generation considered, the sum of add-ons was 1.4¢/kWh.

Renewable technologies

Renewable technologies – hydro, wind, solar thermal, photovoltaics (PV), and biomass – are a diverse group using abundant indigenous resources. In most parts of the world, at least one of the resources is available in sufficient quantity to be an attractive option for future power supply. The technologies are highly modular and benefit from significant economies of mass production. However, until the resource potential is fully understood and appreciated and the value is recognized in decision-making processes, product orders are unlikely to reach the levels where these economies of production can be realized. (See Figure 7.) Activities to accelerate technology readiness and to prepare and expand early markets for these technologies will be critical to their adoption.

Wind technology has matured significantly. The cost of wind-generated electricity is dropping to 5¢/kWh and below in areas where wind speeds are favourable. In many areas of the world, these costs are highly competitive with other technology options, including fossil fuel generation. Targets for wind energy are being established in Europe, and a major programme for

---

[4] A number of approaches were used to derive these figures. These ranged from potential cost of control to assessment of future damages. No universal agreement exists on the methodology used to calculate these costs. They are usually a compromise developed during public hearings.

Figure 6 US states that require or are considering some form of environmental externality factor for discouraging the building of 'polluting' generation

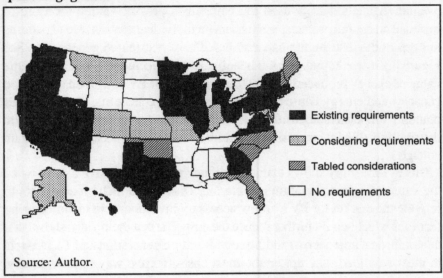

Existing requirements

Considering requirements

Tabled considerations

No requirements

Source: Author.

Figure 7 Renewable costs fell sharply in the 1980s

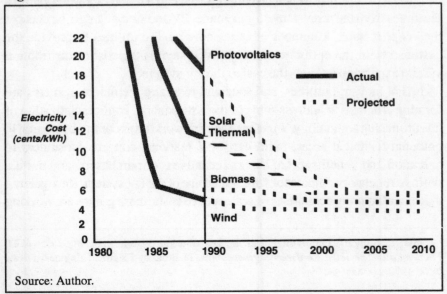

Source: Author.

developing 10,000MW in the US Midwest is being discussed by a number of interested groups.

A PG&E study evaluated the use of PV in the distribution system.[5] The evaluation methodology used the company's present value procedure to evaluate all capital projects, and its data on individual customer load patterns to capture the distribution capacity benefits of distributed generation. (See Figure 8.) It was found that the benefits at the selected location exceed the value of energy produced even at today's costs for PV. (Avoided generation capacity and energy constitute only about half the benefits. The other half comes from distribution benefits such as savings on transmission and distribution (T&D), increased reliability, and power needed to maintain voltage.)

This PG&E study shows that a distributed siting might have almost twice the value of a central station installation. Distributed niche markets could provide the market for PV to allow necessary investments in manufacturing facilities which would further reduce the cost. The evaluation also shows that the traditional approach of adding central station generation and T&D assets to meet new load may not be the most cost-effective way to provide new service. PV, small generators, customer energy efficiency and storage might be lower-cost options. When providing service to a home or a water pump a few miles from the nearest line, for example, PV and storage might be cheaper than copper wire. A number of states now require utilities to provide the customer with the option to install a PV system if the cost of distribution system expansion exceeds the cost of the PV system.

The PV industry, utilities, and state and federal government agencies are forming alliances to address some of the issues facing commercialization of this promising technology. Together, they are working to continue research in potential technical areas, build early PV markets, educate consumers in operation and benefits, establish tax incentives for purchasers, and build a positive regulatory climate for widespread use of the PV system. Recognizing that manufacturing economies of scale reduce costs, these groups are working

[5] Daniel S. Shugar, 'Photovoltaics in the Distribution System', *Twenty First IEEE Photovoltaics Specialist Conference Proceedings*, Institute of Electrical Engineers, New York, 1990, pp. 836–44.

Figure 8 Distributed PV power benefits may exceed current cost in T&D systems, and have twice the value of central PV

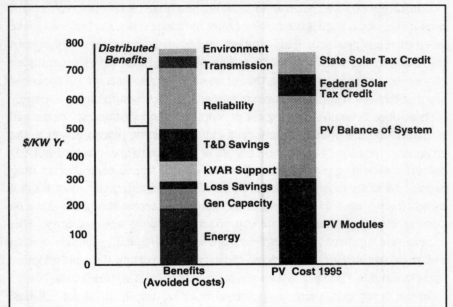

Notes: VAR=Voltage Ampacity Reactive. (Could be equated with turbulence. Involves costs to be handled and overcome.); 1kVAR=1,000VAR.
Source: Author.

to build a sufficiently large market to stimulate manufacturing investments. These efforts should establish the value of the technology, spur early markets, and provide market signals for the emerging PV industry.

Emerging technologies

Emerging technologies may hold the key to resolving such utility issues as a cleaner environment and cost control. Perhaps, more importantly, they might be the instruments for substantial industry change in the way we produce and use electricity. As noted earlier, the old technologies, such as coal and nuclear, relied on economies of scale to obtain cost efficiencies, but in many cases there are no more economies of scale left to capture. The emerging technologies tend to be clean, small and modular, and they achieve their cost efficiencies through economies of product, or mass production. For example, instead of making *bigger* on-site generator sets, wind machines, solar photovoltaics, or high-efficiency compact fluorescent lamps to reduce costs, the factories make *more* of these products. This concept embodies a fundamental change of mindset for the utility industry, which is more familiar with capturing economies in field construction than economies in manufacturing. One could say that this is a shift from 'constructed' energy to 'manufactured' energy.

These new technologies bring many benefits and opportunities. Their small size and modular construction means that they can be placed nearer to the customer. Location of generation near or at the point of use reduces the T&D costs that would be required to deliver the energy from a central generating facility. Modular technologies can be installed in incremental capacities that more closely meet the demand at the time. In a sense, the utility may be coming near to a 'just-in-time' approach to adding new capacity. This incremental approach improves the effectiveness of capital by quickly putting a plant into service with little excess capacity. It also permits the rapid adoption of technological changes, which might be occurring every few years.

On the down side, there are potential negative implications for utilities. Customers can use many of the new modular technologies for self-generation. They could leave the utility system altogether, or worse (from the point of view of utility dispatch and operation), use the utility for back-up only, thereby imposing a cost burden on other customers.

The utility of tomorrow

The utility of tomorrow may evolve into a very new structure that relies on economies of product production and sells unbundled energy services. While many of the old assets will be used for a long time to come, they may play new roles and some may be replaced by new technologies. New growth and service may require new technologies to comply with the business and regulatory conditions of the future.

The emergence of small, clean technologies means that generation sources and energy storage can be placed on the distribution system near the customer. The evolution of this system would substantially change the structure and operation of electric utilities. The future utility structure could be a hybrid of the traditional central and distributed generation and energy efficiency structures. (See Figure 9.) The term 'distributed utility' describes this future, less centralized utility. Its generation units will be dispersed, with small modular technologies used to supply power at crucial points in the distribution system. A wider range of energy resources and technologies will generate power, and the point of generation will tend to be closer to the customer.

Many US utilities have already begun the transformation to the utility of the future. They are not building large plants. Independent power producer markets are expanding, using small generators, including many powered by natural gas and renewable energy technologies. Some customers are cogenerating, and energy efficiency and load management are becoming common practice with the encouragement and support of many utilities, regulators, and policymakers.

Figure 9 Distributed generation, storage and efficiency

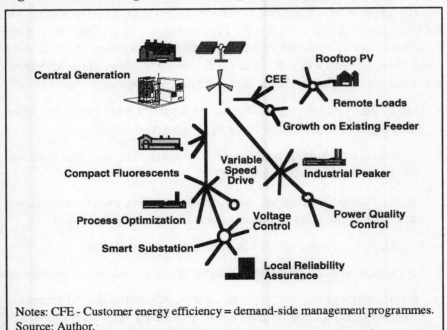

Notes: CFE - Customer energy efficiency = demand-side management programmes.
Source: Author.

The cost of service

The utility of the future will probably need new economic and accounting methods. The traditional cost of service is a function of large central generation (+G), transmission (T), distribution (D), and, for some utilities, large central storage (S), such as pumped hydro. T and D are necessary fixed costs for getting the product to market. The most variable factor was the cost of +G, and utilities therefore focused on minimizing busbar energy costs. This cost of service equation may be written:

$$\$ = f\,(+G,\ S,\ T,\ D)$$

With the decline in improvements in economies of scale and the promise of new technologies, the traditional cost of service equation becomes more complicated. Providing service to new customers does not implicitly mean that the least costly service will be provided by new, large, central station power plants and that bulk power cost or busbar energy cost is the only significant influence on cost of service.

For the utility of the future, competitive service will be provided by minimizing the total cost of service, not just the busbar energy costs. Many options are available to serve a customer. Providing either supply or efficiency close to the customer is, in many cases, the least costly service solution. Some of the new options, and hence factors, in the cost of service equation, are:

- Energy efficiency measures that act on a system wide basis, examples being high efficiency refrigeration or lighting rebates (-G)

- Local small distributed generation, such as small natural-gas-fired generator sets, PV, and perhaps fuel cells (+g)

- Locally directed load management or energy efficiency improvements, say for a large commercial customer or a residential community (-g)

- Local small storage, such as batteries or propane tanks (s)

- Service options tailored to the customers' specific needs and requirements

- 'Service without the wire' options in areas where the cost of providing distribution lines exceeds the cost of generating with hybrid systems (renewables and small generators).

The lowest cost of service that integrates these other factors may be written:

$$\$ = f\ (\pm G,\ S,\ T,\ D,\ \pm g,\ s)$$

The concept of lowest cost of service places the utility in the central role of 'integrator' of numerous and diverse energy service options, and it provides a means for developing better methodologies for balancing supply and demand.

Some new emerging technologies are also altering the equation:

- Repowering options (advanced combustion processes, fuel cells, aero-derivative turbines) for old fossil fuel plants can extend the utilization of prime sites.

- New central station expansion options (wind, solar thermal, compressed air energy storage, and, perhaps, advanced nuclear concepts) are clean and safe and use indigenous resources.

- New techniques (dynamic thermal ratings of transmission cables and transformers) and new technologies (superconductivity) offer improvements for T.

- New technologies in communications, sensors and controls can, and may well be required to, improve the performance of D.

The energy portfolio

Flexibility strengthens sustainability, particularly in situations characterized by uncertainty such as that currently surrounding the utility industry. The prudent utility will develop and maintain a robust portfolio of technologies to manage its future. The composition and relative emphasis of the utilities' portfolios will vary widely. Each utility must identify the portfolio most appropriate for its circumstances. Most will choose from the following options:

- *The 'old' fossil fuel and nuclear option* This option consists of the traditional coal, oil and gas boilers and light water nuclear power plants. It is vulnerable to regulation and lower cost competing technologies. The best management approach would maintain the performance of existing facilities with short payback investments. New, long-term investments in these technologies would be risky. National policies on nuclear power vary greatly. In the United States, it is difficult to site new nuclear power plants. International concern has developed over Russian-designed power plants. Yet advanced design nuclear power plants are being built in several countries.

- *The 'new' fossil fuel option* This option consists of high-efficiency advanced gas turbines and fuel cells fuelled by natural gas. Natural gas appears to be very plentiful, and it is cleaner than other fossil fuels. Development of high-efficiency conversion technologies will enhance the environmental compatibility of technologies using natural gas or gasified biomass and will reduce vulnerability to price increases. In some scenarios, advanced coal conversion might be included, but this resource is vulnerable to regulation and to concerns about global climate change.

- *The renewable option* This option includes hydroelectric, geothermal, wind, solar thermal, PV, and biomass technologies. It is the cleanest generation technology option, but is not without environmental impacts, especially from hydroelectric and geothermal technologies. It is versatile, matches a traditional central station or distributed generation utility structure, relies on indigenous resources, and is the most sustainable of the options. While some of the technologies are cost-effective now, others are not but their costs are declining.

- *The energy efficiency option* This option consists of the hundreds of new, more efficient building and appliance technologies becoming available to utility customers. It might be the least costly and cleanest option. The size of the actual market is unknown, but is probably substantial. Due to the sizeable transaction costs, the utility industry may have to help in bringing this option to its highest potential.

- *The future option* This option consists of passive nuclear, hydrogen, fusion, and other advanced technologies. While offering many potential advantages for the future, it is beyond most utilities' strategic timeframe.

A variety of technical tools must be developed to assist in the task of selecting and maintaining an energy technology investment portfolio. Existing techniques are inadequate to evaluate many new options for utility investment because they do not implicitly include environmental, regulatory, fuel price, and load demand risks. The central challenge is to find a mix of investments that minimizes the overall cost of meeting the ultimate demands of electricity customers (i.e. a reliable source of lighting, space conditioning, mechanical power and other end uses); meets environmental goals; and leads towards a sustainable energy future. This appears closer to stock investment portfolio theories than to present resource expansion models. Such theories are being developed in the United States.

Electricity markets are crucially important if renewables are to play a significant role in meeting world energy needs. Increasing electricity demands are likely to dominate energy demand growth worldwide in spite of electricity's relatively high cost. Electricity makes it possible to apply power only where and when it is needed, and proves to be an ideal energy source for modern factory, commercial and household equipment.

Electricity can be produced from renewable resources in many different ways. The diversity of options means that most parts of the world have a renewable resource that can be used to generate electricity at competitive prices. But diversity frustrates a simple assessment of value. Techniques used to evaluate traditional utility investments do not work well when applied to renewable equipment whose value depends heavily on details of local conditions (local wind or sunlight resources, the hourly pattern of electricity demand in the region, and the characteristics of other equipment operating in the utility system).

Existing evaluation techniques, however, are also inadequate for evaluating many other areas open for utility investment. Future utility portfolios will include investments in thermal generating equipment with characteristics substantially different from those in place today, more dispersed electricity storage systems, advanced transmission and distribution systems, advanced control systems, as well as investments in renewable electricity generation equipment.

Utilities will, therefore, not simply ask whether renewable electricity technology can compete with a particular type of conventional generating plant. They will ask whether investment in renewables makes sense as a part of an efficient electricity production and delivery system.

Assessing the performance of an electricity system that combines equipment with widely varying operating characteristics is a complex analytical task. In a dynamic electricity system, the performance of each item of equipment affects the economics of other devices in the system in complex ways. A balanced analysis must consider such things as:

- How is overall reliability of electricity service affected by different kinds of investment (renewables with intermittent output affect reliability, but reliability also depends on the details of transmission and distribution systems and other generating equipment).

- What is the optimum size for individual generating plants.

- Where should generating equipment and storage systems be located (there are clear advantages in locating generating equipment close to consumption sites since delivering electricity to customers can cost as much as or more than generating electricity).

These are not new issues. But they are receiving renewed attention as utilities struggle to understand the merits of new technologies and to cope with increasing pressure to reduce costs.

New investment opportunities

Utilities today must assemble investment portfolios whose merits cannot easily be assessed using standard methods. Some renewable electricity generators have unique characteristics that complicate investment decisions. While biomass power plants, like fossil fuel power plants, are 'dispatchable' (i.e. their output is under the control of the utility), intermittent renewable systems, such as wind, PV, and solar thermal units, are not. But the characteristics of other potential utility investments (advanced fossil fuel generators, storage, transmission and distribution, customer end use equipment) are also very different from conventional utility investments. New analytical techniques would be needed even if renewables were not available.

Biomass generation equipment, which (typically with a capacity of 100MW or less) would be much smaller than conventional baseload plants, can be dispatched like conventional fossil-fuel-powered equipment. Other renewable systems have more complex operating characteristics. The output of intermittent renewable electricity systems depends on the availability of sunlight or wind. The value of power produced in this way is sensitive to local conditions. Photovoltaic and solar thermal devices located in southern regions where electricity demand is highest during the day can reduce expensive peak generation as well as peak loads on transmission and distribution (T&D) systems.

Hydroelectric systems with large reservoirs can be dispatched to follow fluctuating electricity demands and can help offset variations introduced by intermittent electricity generators operating in the same system. Concern for the environment, and the needs of agriculture and recreational activities, limit the variation possible. Some hydroelectric systems (such as 'run-of-the-river units') cannot be dispatched.

New generating equipment using fossil fuels, however, also requires new ways of thinking about utility investments. The new generation of small gas turbines, for example, can come close to competing with baseload coal plants. The value of small-scale generators increases if a market can be found for the heat left over after electric power is generated, if the equipment is located where it can reduce the investment in transmission and distribution equipment, or if generation systems located close to consumption sites can contribute to system reliability.

Unfortunately, it is particularly difficult to compare investments in T&D with investments in generation and efficiency. Transmission and distribution have too often been considered as an 'overhead' – an inflexible fixed percentage of generation investment – by utility planners. Pressure to minimize costs is, however, forcing many utilities to improve T&D investment and reliability analysis is integrated with other utility investment planning.

These analyses will also be affected by the splitting of the utility industry into separate generation, transmission and distribution functions. A number of functions at present performed by the vertically integrated industry will have to be explicitly described and their value determined. As separate companies, the understanding of how they interact will become ever more important.

Wise investment in T&D is clearly important. This equipment represents 40 per cent of total utility capital investment in the United States and a much higher percentage of all new capital investment in many regions. Investments are lower in densely populated areas (such as northern Europe) and greater in sparsely settled areas. Transmission and distribution systems can also be costly because of energy losses. The amount of electricity generated that is lost in T&D systems averages 6.2 per cent in the United States, and is reported to be much higher in some developing countries. Customer power failures are at least as likely to result from faults in T&D as they are from generation. The cost and reliability of transmission is strongly affected by the location and size of individual generators and storage systems.

Computer networks, telephone switchboards, and many other large systems operating in modern businesses are evolving away from hierarchical models towards more flexible systems whose many nodes have multiple connections, each operating with comparative independence. Utility systems are likely to follow a similar path as they attempt to manage a complex set of demand management and generating equipment distributed through their service areas. At present, most utility grids are operated essentially like an irrigation system – delivering a commodity from a large reservoir to many customer sites. But future utilities are more likely to resemble computer networks, with many sources, many consumers, and continuous re-evaluation of delivery priorities and management of faults. All customers and producers will be able to communicate freely through this system to signal changed priorities and costs. A modern system would, in effect, facilitate a constantly shifting market for electricity producers and consumers.

The development of multiple communication pathways to customers, the so-called 'information highway', is proceeding at a rate faster than that considered realistic by conventional wisdom a few years ago. In November 1993, Pacific Bell and AT&T announced that they will begin construction in 1994 and expect to have 1.5 million Californian homes plugged in by the end of 1996, and more than 5 million by the end of this century. Other US companies are planning the same rapid implementation in other parts of the country. The universal playing of computer games using a television set (Nintendo, Sega, etc.) also means that a new generation of customer is being trained in communication techniques unheard of a decade ago. Utilities need to think and plan carefully how they can utilize these capabilities to their competitive advantage.

The new competitive pressures have forced utilities to behave much more like modern manufacturing enterprises and less like stable state monopolies. Automobile manufacturers, producers of consumer electronics, and many other manufacturing firms have been forced by competition to pay much more attention to the needs of individual clients. They have learned that their survival in a competitive world depends on an ability to understand what it is their customers want, and to react quickly. Often, what customers are looking for is quality. Utilities are quickly learning the same lesson. They have come to understand that customers are not necessarily interested in low-cost kilowatt hours, but instead in low-cost, high-quality energy services.

Quality clearly depends in part on the nature of the services provided by the utility and on the reliability of service. But techniques used to understand how consumers value different levels of reliability, and how different utility investment decisions affect reliability, are inadequate. The need to improve these tools is made all the more urgent by the fact that renewable generating technology and the other new utility investment opportunities can affect system reliability in complex ways. Improved analytical tools for understanding the operation of dynamic utility systems are only a part of the solution. Like other manufacturing enterprises, utilities operating in competitive markets must be sensitive to a wide range of subtle factors that shape consumer decisions.

A portfolio analysis

An attempt was made to analyse a number of investment portfolios. The analysis concentrated on a single simple measure of a portfolio's value: what is the cost of meeting the demand for electricity in the utility's region in the course of a year, assuming that the utility receives a fixed rate of return on its capital investments? This measure forces attention to the complex interactions between different kinds of utility equipment. The basic ground rules are outlined in the Box overleaf. An important simplifying assumption was that all cost-effective investments in energy efficiency and demand management had already been made in order to minimize demand. At present no satisfactory resource expansion models allow energy efficiency to be treated as a supply option, meaning that calculations of efficient investments have to be undertaken in

Ground rules for portfolio analysis

Utility portfolios are assessed by estimating the cost of operating a hypothetical utility during the first quarter of the twenty first century. Data regarding hourly utility loads, wind and sunlight were taken from the PG&E service territory as it existed in 1989. The following assumptions were used in the analysis:

- All cost-effective investments in energy efficiency and demand management have already been made. In many cases such investments would be less expensive than any investment in new generating equipment. The goal of the present analysis is to minimize the cost of meeting the remaining load.

- The utility is free to invest in biomass and conventional generating equipment which minimizes production costs (real utilities, of course, would need to consider the sunk costs of existing equipment).

- Coal prices remain fixed at $2.1 per gigajoule and natural gas prices rise from $4 per gigajoule in the first year to $5 per gigajoule after 30 years. These prices are lower than those used in most long-term forecasts (the US Energy Information Administration, for example, suggests that natural gas prices may approach $8 per gigajoule during this period), because of an assumption that rapid introduction of renewable fuels would increase the diversity of world fuel supplies, thus increasing competition and stabilizing prices.

- The utility is financed assuming both a 6 per cent real cost of capital (typical of regulated utilities) and a 12 per cent real cost of capital (typical of non-regulated owners). A charge of 0.5 per cent for annual insurance is also assumed. No taxes are taken into account in computing fixed charge rates. (Electricity from capital-intensive equipment like renewables is more expensive if high discount rates are used.)

- No allowance is made for environmental costs except that the cost of meeting existing (1992) US environmental regulations is implicit in the price of conventional generating equipment.

- Energy storage equipment is used only for energy stored in hydroelectric reservoirs and in biomass fuels.

- No attempt is made to optimize the mix of wind, solar thermal, and photovoltaic equipment used by the utility.

- Any intermittent renewable output which exceeds the load is wasted. This is not a realistic assumption since a market would undoubtedly be found for this inexpensive energy – however unreliable. Surplus energy could be sold to neighbouring utilities or sold for pumping, desalination, or making other storable products.

- Hydroelectric equipment is dispatched to reduce utility peaks as much as possible within specified constraints on reservoir size and maximum filling and discharge rates.

two steps: the drop in demand due to efficiency has to be calculated and subtracted, and analysis must then concentrate on how to supply the remaining demand. If the supply is more expensive, further reduction in demand is needed. There are no good models which allow two steps to be incorporated into one.

While much work remains to be done, and results must be tested in a large number of different climate and resource regions, the analysis resulted in the following conclusions for a summer peaking utility operating in the first quarter of the twenty-first century (see Figure 10):

- Utility portfolios that rely on intermittent renewable sources for 30 per cent of their electricity can serve loads at lower costs than utilities using typical new equipment and at costs about 5 per cent higher than those achievable using advanced coal and gas production equipment now being developed.

- The cost of the renewable portfolio just described would be unchanged if biomass were substituted for coal. This system would rely on renewables for over 90 per cent of its energy.

- A system meeting 50 per cent of its loads from intermittent renewable resources would cost 10 per cent more than a system using intermittents for 30 per cent of its loads.

- The three conclusions just described assume a 6 per cent real discount rate (typical of regulated utilities in industrialized nations) and equipment costs reasonable for the first quarter of the next century. If a 12 per cent real rate of return is used (or if all capital costs are increased by two-thirds), the cost would be approximately 20 per cent greater.

- Small generating equipment that can be located close to demand centres can reduce transmission and distribution costs and thereby reduce delivered costs. These credits can reduce the gap between cost of a utility with 30 per cent of its power from intermittents and an advanced fossil-based utility to well below 5 per cent (assuming a 6 per cent discount rate).

- Advanced turbine systems using natural gas as a fuel provide an excellent complement to utility portfolios which rely heavily on renewables. They can operate efficiently while following loads and they can be added quickly.

Figure 10 Cost and emission for generation portfolios

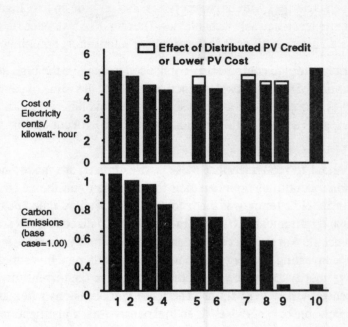

Notes:
1=convention fossil
2=best new fossil
3=advanced fossil
4=advanced fossil with 21% hydro
5=advanced fossil with 21% hydro & 10% PV
6=advanced fossil with 10% mixed intermittents
7=advanced fossil with 30% mixed intermittents (3 wind sites)
8=advanced fossil with 21% hydro & 30% mixed intermittents (3 wind sites)
9=advanced biomass& gas, 21% hydro & 30% mixed intermittents (3 wind sites)
10=Advanced biomass & gas, 21% hydro & 50% mixed intermittents (3 wind sites).

Source: T. Johansson et al. (eds), *Renewable Energy: Sources for Fuels and Electricity*, Island Press, Washington DC, 1993.

Their low cost makes them an excellent way to buy time to determine the best investment strategy in the future.

- Hydroelectric sites with large reservoirs also provide an excellent match for intermittent renewable technologies. The output of hydroelectric systems can be adjusted to fill gaps in the output of intermittent equipment.

- Electricity storage equipment is not needed to achieve the high levels of penetration of renewables just described. In fact, the value of storage to a utility is decreased – not increased – if photovoltaic or solar-thermal equipment is added to utility systems where peaks occur during daylight hours.

It is important to recognize that these cost estimates are made under the assumption that utilities optimize their investments to minimize costs over the long term. No attempt was made to develop a detailed schedule of plant additions and retirements.[6] Utilities in most industrialized nations are likely to experience slow or even declining demand for electricity given the attractive options for investing in energy efficiency. As a result, new investments will be made primarily to replace obsolete equipment or equipment forced into obsolescence by new regulations. There are few occasions when any new plant – renewable or otherwise – is so inexpensive that a justification can be found for replacing equipment which has not reached the end of its useful life. In such cases the full cost of new generating equipment must compete with the operating costs of the existing units. What is becoming clear is that a combination of fossil fuel, renewable and energy efficiency technologies can be devised that provides the lowest-cost option for the long run. Considering only the next increment will not lead to long-term lowest cost nor towards a sustainable energy mix. This poses a dilemma for energy planners. Numerous articles are beginning to appear in the US literature discussing portfolio mixes, usually under the title of Integrated Resource Planning (IRP). The dilemma, in a fundamental sense, is whether short-term market rates should dominate and remove guarantees of long-term capital recovery or, whether long-term lowest cost should dominate and sustain guarantees of long-term capital

---

[6] This might affect costs a little, but not to any significant extent. A schedule without hold-ups was assumed.

recovery. The present regulatory situation, at least in the US, is one where there is a push towards lowest short-term market rates – which favour gas-powered turbine-based power plants – while still guaranteeing long-term capital recovery.[7] Thus it will be difficult – if not impossible – to provide a more diversified, sustainable energy mix. It also emphasizes the importance of this issue since investments made today will have the potential to preclude options in the next 30 years.

Towards a portfolio for a sustainable energy future

Utilities will be making decisions about their energy technology portfolios in concert with the regulators and policymakers who influence their business. A great deal of uncertainty remains about the evolution of this process. How will people decide what to do? How do we guide investments towards the correct energy portfolio?

Countries in the process of building utility systems may be in a better position to exploit new technologies than those with large investments in old equipment. Industrialized nations with large sunk costs may be tempted by investments which make near-term sense but are less suitable for future markets. Commitment to large generating plants having 30-year lifespans can create a self-fulfilling prophecy by making investments in new technologies uneconomical for a generation to come. Emerging technologies allow utilities to achieve ambitious environmental goals while keeping the costs of energy services low and the quality high. However, the regulatory and policy environment governing utility decisions in most regions has evolved from an antiquated paradigm of utility technology and structure and, unless changed, is likely to frustrate the emergence of utility portfolios that make optimum use of renewable and other emerging technologies. This may be particularly

---

[7] Guarantees are provided either by regulators or by government. They were appropriate for almost 80 years and there is now a large residue of inertia. In the US, this manifests itself in one way as a fuel adjustment clause, allowing utilities automatically to recover any increased costs of fuel. Hence, they take no long-term fuel price risk. Further, almost all contracts for power purchases are for a term of 30 years, so constituting a take-or-pay situation. If short term rates are really the goal, why not let the market decide the terms? The independent power producer would then take a bigger risk and would have to charge more, which would drive up short term rates. All of these considerations are currently in a state of flux.

important for developing countries which do not have well-established utility systems. They may find it advantageous to start with smaller village power systems which include renewable resources that then coalesce into a larger system. The alternative of starting with large power plants will involve overwhelming T&D costs in bringing electricity to the villages. The situation necessitates new approaches to public policy and regulation to encourage the transition to utility businesses better able to meet the costs, quality, flexibility and environmental requirements of future electricity markets. The role of the 'portfolio manager' is thus an important one. Who will take this position in an industry divided into its component functions of generation, transmission and distribution will be a major problem.

Deliberations regarding portfolio investment criteria must consider the perspectives both of business and of society. Government has a legitimate, albeit limited, role in these deliberations. These new analytic tools are also likely to incorporate techniques which address environmental externalities. In fact, this issue is already being addressed. Following New York's lead, over 75 per cent of states in the US now require, or are considering, incorporation of environmental externality add-ons to discourage 'polluting' generation. US state regulators are addressing this issue by a wide variety of methods in three regulatory arenas:

* *Planning or acquiring new resources* Three accounting approaches are being used to modify resource planning and acquisition procedures to better reflect environmental externalities. The first gives qualitative ratings to the various demand- and supply-side energy technology options, with higher ratings for the environmentally preferred options. The second approach adds or deducts a percentage of the value of the resource, depending on its relative environmental impact.[8] The final approach estimates and adds the actual cost of the externality to the comparative evaluation of resource alternatives. In each case, but particularly in the final instance, debates ensue regarding the relative impacts of resources, technologies, and absolute values of externality estimates.

---

[8] This is a way of trying to incorporate some of the externalities. For example, some states make the assumption that they are willing to pay 10 per cent more for renewable energy than for conventional sources, so if bids are within that range they win. Some require coal to be at least 10 per cent below any other bid to win.

- *Profit incentives* Some states are allowing rate incentives for low-polluting generation and energy efficiency programmes, with environmental externalities providing the rationale. This allows the utility to earn a higher rate of return and subsequently make higher profits.

- *Set asides* Some states also require that a specified amount of energy efficiency and renewable energy technologies be procured as part of an overall new resource programme.

In a recent evaluation, Clem Sunter, head of scenario planning for a major corporation, concluded that 'government must perform a limited role competently'.[9] It is imperative that citizens become actively involved in this public process because it bears directly on setting the criteria for portfolio selection and investment and establishing a basis for assessing the value of the technology options and the overall cost of energy. In the predominant cases in the US where public interest and environmental groups have become involved in integrated resource planning (portfolio selection) the outcome has been a shift towards greater use of energy efficiency and renewable technologies, i.e. towards sustainability. The general public seem to be ahead of energy policy decisionmakers.

In a recent book, Jane Jacobs postulates the existence of two parallel moral syndromes – one commercial, the other acting as a guardian – in which individuals and organizations operate.[10] She concludes that problems arise when an organization or the people who constitute it, attempt to operate outside the syndrome appropriate to its activity. The government and the commercial sectors should pursue goals and functions which are appropriate to their culture and ethics, and should avoid activities requiring operations fundamentally beyond the bounds of their cultural and ethical context. Her conclusions echoed Sunter's: 'Government best serves the function of supporting commerce, preventing its excesses, and providing incentives for commerce to develop in sectors deemed to be in the common good.' And once the directions are set, industry should provide what a society wants by making a profit in a competitive environment.

---

[9] Clem Sunter, *The New Century: Quest for the High Road*, Human and Rousseau, Tafelberg, South Africa, 1992.
[10] Jane Jacobs, *Systems of Survival: A Dialogue on the Moral Foundations of Commerce and Politics*, Random House, New York, 1992.

Government still plays an important role. In a recent interview, Gro Harlem Brundtland, chair of the 1987 United Nations World Commission on Environment and Development and now Prime Minister of Norway, reflected on the requirements for moving towards sustainability and noted that 'such changes will require political direction and strong support of a democratic populace. Global change depends on our ability to adapt market mechanisms to promote sustainable development.'[11]

Setting investment objectives for a utility portfolio is a political process, and one which can introduce a new value system of sustainability. This is particularly important in the utility sector, as technology investment decisions made today will still be part of the operating utility system thirty or, perhaps, fifty years from now. Today's decisions must look out significantly far into the future. It is equally important that market mechanisms be used to select the technologies or options to fulfil the investment objectives. Policies and regulations should define the objectives set by the political process and allow market forces to build a path to a transitional and, eventually, a sustainable portfolio.

We should not be parochial in our perspective as we review our options. It is important to recognize the importance of the phrase 'all the world's people' in the United Nations World Commissions' conclusions about sustainability.[12] Sunter coined terms to describe the dichotomy present in the world's distribution of wealth and relative standard of living – the 'rich young millions' and the 'poor young billions'. He writes:[13]

> Why worry about the imbalance in the world's income? Apart from the perfectly decent reason of wanting to make the world a better place, one has to be apprehensive about what a high ratio of 50:1 (income differential) could do to global stability in the next century.'

If renewable energy and other emerging environmentally preferred technologies are not a significant part of the global energy base twenty years from now, this will be the result of a failure, not of technology, but of society's

---

[11] 'The Road from Rio: An Interview with Gro Harlem Brundtland', *Technology Review*, April 1993, pp. 61–5.
[12] UNWCED, *Our Common Future*, op. cit.
[13] Sunter, op. cit.

vision to make prudent investments in a portfolio targeting a sustainable energy future. Those pushing towards enoughness and sustainability are accused of knowing the value of everything but the price of nothing. Those wishing to remain with consumption and exploitation are accused of knowing the price of everything but the value of nothing. We need to find a path that considers both value and price.

The political process of establishing energy technology investment portfolios must integrate all parties with an interest in charting a path to sustainability. Everyone must play a role. It is time to move beyond summits, beyond statements, and beyond posturing. For the sake of present and future generations, it is time to make a commitment to action.

Selected bibliography

Shimon Awerbuch, 'The Surprising Role of Risk in Utility Integrated Resource Planning', *Electricity Journal*, Vol. 6 (3), April 1993.

Richard Hirsch, *Technology and Transformation in the American Electric Utility Industry*, Cambridge University Press, Cambridge, 1989.

Thomas B. Johansson, Henry Kelly, Amulya K.N. Reddy, and Robert H. Williams (eds), *Renewable Energy: Sources for Fuels and Electricity*, Island Press, Washington DC, 1993.

P. Joskow and R. Schmalensee, *Markets for Power*, MIT Press, Cambridge, MA, 1985.

Carl J. Weinberg, 'Adding Renewables to the Portfolio', *Electricity Journal*, April 1992, pp. 62–4.

Carl J. Weinberg, 'Enoughness and Sustainability', *Solar Today*, May–June 1992, p. 19.

Carl J. Weinberg and Merwin L. Brown, 'The Electric Utility Industry Trends; A Strategic Response', presented at the Eletrobras Seminar on Profile of the New Utility, Brasilia, 1 October 1992.

Acknowledgements

The author expresses his grateful appreciation to Katie McCormack and Merwin Brown of Pacific Gas and Electric Company, Department of Research and Development. Their input, advice and comments were invaluable in the preparation of this paper.

Development and Transfer of Sustainable Energy Technologies: Opportunities and Impediments

Katsuo Seiki and Yutaka Tsuchida

Abstract

The international policy issue given greatest priority since the UNCED meeting in 1992 has been sustainable development, but there is a wide gap in opinion concerning how sustainable development can be realized. As an alternative to the two traditional responses to the problem – that sustainable development can be pursued either through changes in social systems or through technological solutions – we explore a third approach: realization of sustainable development through changes in social systems and technological solutions.

In light of this concept, technology transfer is one of the most urgent issues. However, since technologies to be transferred are owned by private businesses, there need to be market mechanisms to motivate the businesses to push for technology transfer.

More detailed examination shows that technology transfer will remain unsolved if it is dependent solely on existing market mechanisms. Our main argument in this paper is that to make technology transfer possible recipient demand must be stimulated, incentive and capacity of the supply side be increased, and an innovative intermediary be established to facilitate contact between the supply side and the recipients. Elements of this are illustrated with reference to the Japanese Green Aid Plan, which includes joint R&D programmes and targeted clean-ups of specific regions to act as both stimulus and demonstration.

In July 1993, the World Resources Institute (WRI) and the Global Industrial and Social Progress Research Institute (GISPRI) proposed a number of action programmes based on these ideas. The programmes called for the harmonization of environmental standards on an international scale, the creation of a network among supply-side businesses, and the establishment of an intermediary organization to provide information on financing and technologies.

However, many basic problems related to achieving technology transfer remain unsolved. To find the solutions, a forum of representatives from the government, non-governmental organizations, and industries should be created to delve into and discuss the issues.

1. Changes in environmental issues and policy development

Current environmental issues are characterized by an expansion of human activity which is exceeding the capacity of the global and regional environment and exerting an irreversible effect on the global and regional ecosystem.

Accordingly, resolution of these issues should aim not at responding to each individual incidence of an environmental problem but at changing the system whereby mass production leads to consumption and then to waste. In this sense, current environmental issues have a wider importance, involving the entire civilization and social structures – issues in which environmental, economic and energy matters must be addressed together.

In response to this change in the character of environmental issues, environmental policy in the 1990s is also undergoing new developments. The Organization for Economic Cooperation and Development (OECD) identified sustainable development, long-term strategic planning, product lifecycle analysis, integrated pollution prevention and control and voluntary agreement as factors to be addressed in environmental policy in the 1990s. This reveals a great contrast with the factors outlined in the 1970s which included limits to growth, pollution clean-up and liability. (See Figure 1.)

The problem is the concept behind the term 'sustainable development'. While there is general agreement that 'sustainable development' is the key term in these policies, opinion is divided on how to achieve it. Theorists such as Professor D.L. Meadows of New Hampshire University believe that these issues can only be solved by restraining aspiration while those such as Dr F. Fukuyama of the RAND Corporation emphasize promoting technological development based on democracy and a market economy. (See Box 1.) These two views are more or less extreme ones, representing respectively so-called optimism in human nature and technological optimism.

2. The need for technological resolutions

We should probably take a different approach – a third road – by forming a new social system and advancing technological development in tandem. (Obviously, the nature of the technology which is developed exerts an impact on the social system.)

Figure 1 The flow of environmental conservation measures in OECD

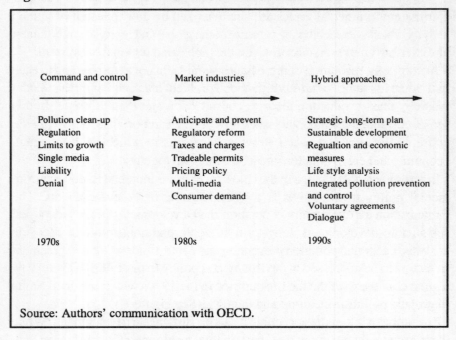

Command and control	Market industries	Hybrid approaches
Pollution clean-up	Anticipate and prevent	Strategic long-term plan
Regulation	Regulatory reform	Sustainable development
Limits to growth	Taxes and charges	Regualtion and economic
Single media	Tradeable permits	measures
Liability	Pricing policy	Life style analysis
Denial	Multi-media	Integrated pollution prevention
	Consumer demand	and control
		Voluntary agreements
		Dialogue
1970s	1980s	1990s

Source: Authors' communication with OECD.

In particular, in the case of developing countries, high aspirations towards for economic development, combined with population explosion have made it extremely difficult to control environmental problems such as increasing CO_2 emissions. So, as is shown by the Kaya equation (See Box 2), there is a need for energy conservation and decarbonization policies in order to maintain economic expansions while simultaneously protecting the environment.[1] A new energy-efficient model of development is needed for developing countries in place of the development models adopted by Japan and Korea, focusing on heavy industry.

---

[1] Y. Kaya, K. Miyaji, and R. Matsuhashi, 'A Grand Strategy for Global Warming', paper presented at the Tokyo Conference on 'The Global Environment and Human Response towards Sustainable Development', September 1989. Y. Ogawa, 'Factor Analysis of Regional and Sectoral Differences in Energy Consumptions and CO_2 Emissions – An Example of Thematic Assessments', Paper presented at the meeting of Energy Industry Subgroup (EIS), Working Group 3 of IPCC, 12-13 November 1991, Seoul.

Some of the problems of working towards sustainable development are particularly apparent in Asian countries with high growth rates. In East Asia, soil degradation is said to have already reached a level three to eight times higher than the world's average. Carbon dioxide emissions per unit of GDP are three times larger than those of Latin America, and energy consumption per unit of GNP is nearly twice as high. Furthermore, in the East Asian region, excluding the People's Republic of China, as much as 2 million hectares of forests, or 0.7 per cent of the total forest area, are disappearing from the earth's surface each year. (See Figure 2.)[2] What is most striking in these regions is the inefficient use of energy and the high rate of CO_2 emissions. Soil and water pollution are also severe, and if these conditions remain unchanged the World Bank estimates that by the year 2000 it will take $20 billion to maintain Asia's environment. This region requires immediate technology transfer.

Recently one of the authors of this paper, Katsuo Seiki, studied the potential for technology transfer to China, and had the following three impressions:

- First, regulations against traditional pollutants, particularly SO_x and NO_x, are sufficiently stringent and there is a great demand for the introduction of pollution-control equipment. However, because of a shortage of capital, there is a growing need to promote technology transfer at the lowest possible cost.

- Second, there is practically no incentive for energy conservation or recycling technology because China's pricing system does not reflect actual market costs. To promote technology transfer in this field, it is important to apply a market economy based pricing scheme.

- Third, China has a strong long-term desire to build up its industry dealing with environmental problems, both that producing pollution control equipment and that producing energy-efficient and new technologies. This will probably constitute the main element of environmental policy because in this way China will be able to acquire necessary technologies and develop eco-industries which will contribute to environmental protection as well

---

[2] World Bank, *World Development Report 1992*, Oxford University Press, New York, 1992.

Box 1 Optimism: human nature versus technology

Dennis L. Meadows (Professor, Department of Business Economics, New Hampshire University: abstract from lecture titled 'Dimensions of the coming global collapse' in the Symposium on 'The Last Cross-Roads', Nagoya, Japan, sponsored by Chukyo Television, 1993).

'The world's population is increasing at a rate never seen before in the history of mankind, and has already exceeded the capacity of the Earth to sustain it. There are other areas where limits have already been exceeded, such as the depletion of fish resources, the ozone layer, arable land, tropical rain forests, and underground water supplies. Since the 1980s, GNP in 60 countries or regions has seen negative growth on a per capita basis. 94 countries suffered negative growth in per capita food production between 1985 and 1990. Using simulation models, we were able to clearly discern the inter-relationships between population increase and economic growth. We discovered that after the year 2000 the world's production will decrease, and after the year 2040 economic growth will stop. Because of mankind's wasteful use of resources, the globe's resources will be exhausted in the year 2050. If we are to find civilized solutions while continuing the growth, there must be restraint of the rise in population and we must do away with our desire to expand production. The problem cannot be solved by scientific and technological extensions of physical limits.'

Francis Fukuyama (Consultant in International Political Science to the RAND Corporation in the US: abstract from lecture titled 'Can the "End of History" save mankind?' in the Symposium on 'The Last Cross-Roads', Nagoya, Japan, sponsored by Chukyo Television, 1993).

'There is no foundation for theories postulating the extinction of mankind or the downfall of civilization, since these theories do not take into account the development of future technologies. Currently there are three approaches being considered as solutions for environmental problems: de-industrialization, zero growth, or the continuation of industrialization.

De-industrialization is not feasible. It is only people in rich countries who can suggest we go back to nature. If society as a whole were to go back to nature, the end result would only be poverty, sickness, and restrictions on women's freedom. In other words, another Somalia.

Zero growth is even less realistic than de-industrialization for the following four reasons:

i Industry is at a high level of advancement, and cannot be pushed backward to lower levels. Pollution and energy problems can only be solved with future technologies.

ii To realize zero growth, it would be necessary to redistribute a country's domestic accumulated wealth. There is however no way to do this democratically.

iii To realize zero growth, it would also be necessary to redistribute the accumulated wealth of countries internationally. It is however impossible to do this simultaneously throughout the world and could very well bring about global warfare.

iv Trying to force people to do what they don't want to do would hinder individual freedoms.

The only realistic way to solve environmental problems is to continue industrialization while expanding economic growth based on technological innovations.

We must emphasize technological developments that can solve environmental problems. The system to solve environmental problems should be one which can solve human problems. Principles based on freedom and democracy, which can ensure a market economy, are the best.'

Box 2 Factors affecting total CO_2 emissions

With CO_2 emissions, carbon per unit of energy, energy intensity, per capita GDP and population described as C, U, S, G, and P respectively,

$$C = U \cdot S \cdot G \cdot P = (CO_2/E) \cdot (E/GDP) \cdot (GDP/P) \cdot P$$

where E is the total energy demand. From this equation, changes in CO_2 emission, dC, can be described as follows:

$$
\begin{aligned}
dC = \ & (C/U) \cdot dU && \text{Rate of change in carbon intensity of energy supply} \\
+ \ & (C/S) \cdot dS && \text{Energy intensity factor} \\
+ \ & (C/G) \cdot dG && \text{Per capita economic growth} \\
+ \ & (C/P) \cdot dP && \text{Population growth}
\end{aligned}
$$

If population (P) and per capita GNP (G) continue to grow, U and S have to decrease in order to maintain dC constant.

---

as economic development. The Chinese government recently established a special economic zone near the Yangtze River with a view to inviting foreign investment in environmental issues.

The study concluded that whether rapid transfer of environmental technology to China does or does not proceed depends on the extent to which enterprises are able to use their limited capital resources for environment protection measures. In this regard, new ideas must be worked out at the levels of both international assistance and the domestic economy.

International trends in energy–environmental technology cooperation

As we have seen, developing countries are confronting the problem of the dilemma between economic development and environmental protection. To solve this problem, the transfer of environmental technology is urgently required.

Figure 2 East Asia: environmental indicators

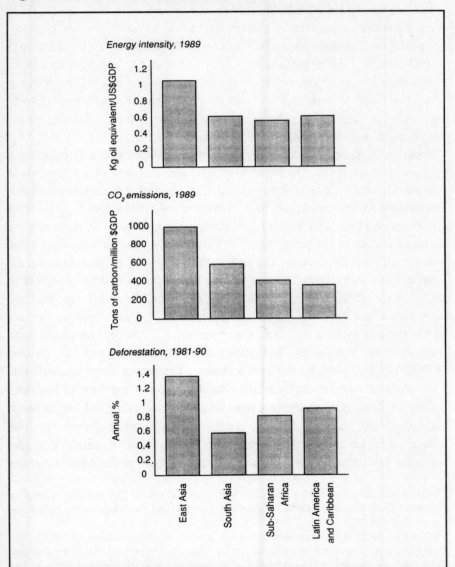

Source: The World Bank, *Sustaining Rapid Development in East Asia and the Pacific*, 'Development in Practice' Series, The World Bank, Washington DC, 1993; World Resources Institute, Washington DC.

The transfer of environmental technology was one of the highest-priority issues discussed at the Earth Summit in Rio de Janeiro in June 1992. While there have been important achievements in the area of capacity building cooperation, the discussions in Rio consisted primarily of conventional North–South style negotiations which focused on 'assured access to technology', 'non-commercial and preferential conditions for transferring technology' and 'compulsory acquisition of intellectual property rights'. New approaches which aim at pragmatic solutions on the basis of North–South cooperation were not advanced. (See Agenda 21.)

Meanwhile, during the Summit, industry played a central role in making proposals which revealed a more pragmatic stance. One of these proposals was embodied in *Changing Course*, drafted by the Business Council for Sustainable Development (BCSD).[3] This proposed that technology transfer can be most effectively carried out when it is undertaken in tandem with building human resource capacity – by training – through long-term joint ventures in the private sector. The publication states that governments should support these undertakings through official development assistance and other forms of aid. (Note that, in the BCSD publication, the words 'technology cooperation' are used in place of 'technology transfer' as the process should be considered to be a two-way communication.) Another proposal from industry is the Sustainable Technology and Energy Efficiency Programme (STEEP) put forward by the International Chamber of Commerce (ICC). This proposal is an energy-environmental version of the Eureka Project,[4] calling for 50 per cent government subsidization for technological development and technology transfer related to energy efficiency and environment. The special characteristic of this programme is its emphasis on introduction and spread of new technologies.[5] These ICC and BCSD proposals both recognize

---

[3] Stephan Schmidheiny with the Business Council for Sustainable Development, *Changing Course: A Global Business Perspective on Development and the Environment*, MIT Press Cambridge, MA, 1992.

[4] EUREKA (the European Research Coordination Agency) was established in 1985 by the European Community to improve the productivity and competitiveness of European industries and economies through closer cooperation of business and research institutes in the area of advance technologies.

[5] International Chamber of Commerce, 'ICC World Industry Council for the Environment-Background Documents for the Versailles Inaugural Meeting', ICC, Paris, 26 February 1993.

that private sector industry should take initiatives directly, and go beyond simple one-shot technology transfer towards cooperation with recipient nations on a long-term basis.

In addition to these two proposals, various information systems have begun to be established as practical steps to promote technology cooperation by providing information on, for example, the characteristics, availability and suppliers of technology. These systems include the International Energy Agency's Technology Information Exchange System (TIES) (where the IEA becomes the key station and technological information clearing house) and the UN Environment Programme, Industry and Environment Office (UNEP-IEO) initiative.

One example of a unique measure for promoting technology transfer is the Green Aid Plan which is being implemented by the Ministry of International Trade and Industry of the Japanese government. Under this plan, the donor and recipient governments engage in policy dialogue through which specific energy-environmental technology cooperation needs are prioritized, and identified technology cooperation together with capacity building and financial cooperation is implemented in a comprehensive manner. One of the features of this programme is the proposal for joint research on and development of technology which conforms to local needs. Local needs are determined by factors such as cost effectiveness, resource endowments and infrastructure.

One example of such a joint R&D project is currently being advanced between the People's Republic of China and Japan. As desulphurization technologies used in advanced nations are too expensive to be used by developing nations, there is a need to develop simple desulphurization facilities which are more reasonable in terms of cost even at the expense of a certain degree of efficiency. This project aims to achieve more than 50 per cent desulphurization efficiency in return for not more than half the cost of a typical standard facility in Japan. Other examples of joint R&D projects are cooperation between Japan and Egypt, and between Japan and Thailand, respectively for waste water and solar power generation technology. Again, specific local conditions are taken into account.

Another aim of these programmes is to enhance the recipient nation's research and development capacity through the cooperation of researchers and engineers. Unless such capacity is developed in these countries, the conditions of technological subordination and 'technological colonialism' will not disappear. In light of these potentially positive effects, we believe that such R&D projects will be important factors in future technology cooperation programmes.

A second unique aspect of the Green Aid Plan is the 'Eco-phoenix Programme'. This is designed to undertake a comprehensive clean-up of a specific region (for example Tianjin in China). Improving the environmental situation of a specific region will lead to the following possibilities:

*   Achieving comprehensive planning and other forms of capacity building;
*   Reaching suitable environmental conservation measures through a systemic approach;
*   Demonstration effects for regions in which technology transfer and environmental conservation are not being set into motion.

These programmes are also likely to become models for future technology cooperation programmes.

Trade and environment

Trade and environment is one of the major themes of the post-Uruguay Round talks, and the international debate on technology cooperation cannot be overlooked.[6]

In connection with the North American Free Trade Agreement (NAFTA) there is much debate in the United States about environmental issues. This debate is not only about the flow of pollutants across borders; it also concerns whether or not countervailing duties and other cross-border measures should be devised to counteract the US importation of low-priced goods which are being produced under less stringent environmental regulations.

Of course, there are cases where environmental measures constitute actual trade impediments. However questioning whether or not the burden of these environmental costs should be readjusted – for example, through subsidies and grants for products made using clean processes – and whether or not any harmonization of environmental standards should be made has major impli-

---

[6] A fuller overview of trade issues is given in the paper by Vincent Cable above.

cations for the issue of technology transfer:[7] any strengthening of regulations in developing countries will lead to an increase in demand for technology transfer. Environmentalists and supporters of free trade are divided in their views on this issue, making it necessary to wait for further debate in the future. For example, Repetto argues that a win-win strategy should be adopted through which a balance could be achieved between trade concessions and strengthening of environmental regulations, whereas Bhagwati argues strongly for support of free trade.[8]

Essential issues related to energy-environmental technology transfer

As already mentioned, despite the fact that energy-environmental technology transfer was named at UNCED as a most urgent issue, no significant progress has as yet been made. There are many actors (governments, industries, NGOs) involved in energy-environmental technology cooperation. However, in such promotion official development assistance funds and other government-based initiatives play only a peripheral, facilitating role. The key players are industry which possesses the technology that needs to be transferred, and the recipient corporations which utilize this technology. This simple fact should not be overlooked. In a market economy, a government cannot compel corporations to act. Therefore, unless industry is motivated to transfer the technology it possesses, there will be no transfer.

There are, however, complications in technology cooperation that do not exist in ordinary commercial trade. The supply side obviously looks for good profit opportunities, so technology transfers which tend to produce relatively lower profits are likely to be given low priority. On the other hand, the demand side has strong development needs and therefore, in general, does not place great priority on improving the environment and increasing energy efficiency.

---

[7] These would not necessarily be aimed at the identical standards in developed and developing countries. For example, minimum levels of production standards could be set.

[8] Robert Repetto, 'Trade and Environment Policies: Achieving Complementaries and Avoiding Conflicts', *Issues and Ideas*, World Resources Institute, Washington, DC, July 1993. J.N. Bhagwati, 'Trade and the Environment: the False Conflict?', in D. Zaelke, P. Orbuch and R.F. Housman (eds), *Trade and the Environment: Law, Economics and Policy*, Island Press, Washington DC, 1993, pp. 159–190.

In developing countries the major factors in reducing demand for these technologies are low energy prices, lax environmental standards, and lack of observance of strict regulations. In China, despite environmental regulations, the actual penalty system is lenient and provides an easy escape, making the regulations ineffective.

In the 1960s and 1970s, Japan experienced serious SO_x and NO_x pollution, but within ten years it has been able completely to resolve SO_x pollution, and, although there are still problems with emissions from mobile sources, overall pollution from NO_x has greatly decreased. The mechanism for achieving this was a combination of three measures:

- The establishment of appropriate regulations
- The promotion of technological development
- Financial assistance for capital investment which enable regulations to be observed.

This is applicable to environmental conservation in developing countries. The building of this three-part mechanism by developing countries is the most important factor in making technology cooperation effective.

Meanwhile, there is also a need to increase the incentives and reduce risks on the supply side. In addition to the provision of soft loans by the public sector, it is necessary to organize industry and create information exchange through groups such as the Industry Cooperative for Ozone Layer Protection (ICOLP).

It is also important that third party intermediaries facilitate contact between the relevant parties in areas such as finance and technological information.

Since private industries play a key role in technology transfer it is better to entrust it to market forces. However, due to the character of technology transfer, it is not sufficient merely to depend on existing markets. Measures to reinforce market functions – that is, to expand demand on the recipient side, to develop supply capacity on the supply side, and to strengthen intermediary functions for making contacts between the two sides – are important in promoting technology transfer. These are the points we wish to emphasize most strongly in this paper.

What responses meet the needs of each type of environmental issue?

Current environmental issues are classified into local, cross-border and global environmental issues.

Global environmental problems including CO_2 emissions – the most serious problem – have the same effect no matter where they originate. Even if one country restricts its emissions of CO_2, an increase of emissions from others can easily cancel out these efforts. However, the situation in developing countries, especially countries with large CO_2 emissions, means that there are currently no incentives to respond to these issues. (This may not be the case for responses which have other aims such as energy conservation.)

These issues must be dealt with urgently. The scheme needed in order to increase the demand for technology transfer and expand supply capacity is referred to as 'joint implementation'. It is based on the fact that no matter where CO_2 is emitted the effect is the same, and that it is most efficient to mitigate emissions in the most cost-effective areas. In the current UN Framework Convention on Climate Change, the targets have not yet been clearly set, making more detailed technological investigations necessary for joint implementation to be effective. However, even partial application of the concept of joint implementation is expected to contribute to increased technology transfer. Projects suitable for joint implementation are those that are 'win-win', benefiting demand side and supply side simultaneously. Whether finances from the World Bank Global Environmental Fund (GEF) and similar funds should be used for promoting such schemes ought to be considered.

Generally speaking, the so-called 'polluter pays' principle should be adopted for SO_x emissions and other cross-border environmental issues. Allocating responsibility is somewhat less clear-cut than in simple local environment pollution cases. While the serious damage caused by acid rain makes an urgent technological response necessary, the problematic emissions transcend borders, considerably diminishing the incentive to control emissions. (Of course, the transboundary pollution can also be a problem in the country where it originates: in this sense there is something of an incentive to restrict emissions.) Therefore it would be most effective to create regional cooperative agreements, and establish common targets for the region (as seen in the UNECE Convention on Long Range Transboundary Air Pollution), thereby forming a partnership for the promotion of technology cooperation.

Particles, NO_x emissions and other local environmental problems are the responsibility of the country which emits them, and should be resolved through its own efforts. However, a sharp worsening of the local environmental pollution could lead to economic collapse and prevention of sustainable development in some developing countries. This could lead to a refugee problem, thereby further aggravating the North–South divide.

In this sense, local environmental conservation is indirectly a global issue, and there is thus a need to go beyond the traditional aid concept and make the resolution of technological issues a priority. Again, while demand on the recipient side might exist, ways must be found to increase both the incentive to supply and capacity of the supply side.

Which technologies should be given priority?

While there is obviously a strong demand at present in developing countries for desulphurization facilities and similar downstream technologies, the long-term solution is to pursue new types of development, giving priority to the following:

*   *Energy and resource conservation technologies* These are the fundamental consideration. Currently, many developing countries use relatively inexpensive energy-intensive technologies. In certain developing countries, such as China, energy costs are low, making the incentive to introduce energy conserving technologies ineffective. However, even in these countries, the cost of energy is rising each year. This, combined with further application of market mechanisms such as the introduction of cost concepts (for example, allowing for depreciation) is expected to increase future demand for these technologies.

*   *Moving from downstream to upstream technologies* As we have seen, technologies for improving the industrial structures and production systems of the developing economies should probably be given priority. In this regard, Seiki encountered the strong desire not only to conserve energy but also to transform the industrial structure and introduce efficient production systems on a study mission in Chongqing in China. This illustrates the importance of taking into consideration the specific economic and industrial structures of the recipient country.

- *Moving from individual component technologies to systemic technologies* Systemic technologies such as 'heat cascading'[9] are receiving increased attention in the field of energy technology. Generally speaking, systemic technologies should be more widely used in energy-environmental technology transfer. Past experience of technology cooperation in developing countries, in fields other than the environmental one, has shown that it is not efficient to transfer only individual technologies: for this reason it is increasingly necessary to introduce systemic technologies. In particular, the need to provide total systems which adjust city planning and infrastructure planning is growing in importance.

There are various theories on 'appropriate technology'. The term does not refer to low-cost environmental technology which is less efficient than advanced technology. In the long term, it entails the most efficient response to local conditions, including resource endowments and technology structure. It also refers to the demand for systemic adjustment and infrastructure planning as mentioned above. Currently, the only way to identify the appropriate technology is to rely on the bottom-up approach (needing case by case determination), but it is important to form a clearer concept of appropriate technology by accumulating various case examples.

Several concrete responses

As we have seen, in order to promote technology cooperation it is necessary (a) to expand the incentive and capacity of the supply side; (b) to increase the demand for energy-environmental technology on the recipient side; and (c) to establish intermediary organizations for linking supply and demand. In this context, we wish to list some measures which can be adopted immediately.[10] Of course, these proposals do not cover all measures for promoting energy-environmental technology cooperation. This list includes examples of measures which at present are thought to be effective.

---

[9] Heat cascading refers to systems in which the heat output from one industrial process is used as the heat input to another (lower-temperature) process – often using the 'waste' heat ultimately to heat buildings.

[10] WRI-GISPRI, 'Joint Statement of the World Resources Institute and the Global Industrial and Social Progress Research Institute', Washington DC, July 1993.

* *Promotion of voluntary international standards for environmental management and technology cooperation* An effective approach for increasing the demand for technology transfer on the recipient side is to promote international standards for environmental protection. Even if harmonization of environmental standards and emission regulations is impossible at this stage, harmonization is one available option for other standards such as environmental management. There is a critical need to continue the debate on this subject. Currently, harmonization of environmental management standards, which was begun by the International Organization for Standardization (ISO), is an important option. In addition, there are policies such as the Charter of Keidanren which have determined informal rules that Japanese companies must operate in foreign countries according to standards established in Japan.[11] By extension, more general standards for technology cooperation could be investigated by ISO and similar organizations.

  Promoting harmonization of standards will tend to lead to standardized technologies, while recipient countries often need more site-specific technologies that reflect local conditions. Achieving the balance between standardized and site-specific technologies will be a task to be addressed in the future.

* *Expansion of technology cooperation via networks of globalizing enterprises* The formation of ICOLP in response to ozone layer depletion has been widely viewed as a useful step in rapidly promoting technology transfer. The creation of fora for exchanging information on practical experience among globalizing firms is thought to have a strong effect on reducing risk and identifying profit opportunities. Currently, these networks are not limited to those in the field of ozone layer depletion. The International Chamber of Commerce World Industry Council on Environment group (ICC-WICE), the Electricity 7 (E7), the World Engineering Congress for Sustainable Development (WECSD) and other organizations have also been formed, and the extension of networks in many other fields should be sought.

---

[11] Keidanren (Japanese Federation of Economic Organizations), 'Keidanren Global Environmental Charter', Tokyo, 23 April 1991.

- *Establishment of innovative intermediary support for technology cooperation* The attempt to build up international links between suppliers and recipients of technology transfer is making very slow progress. The process is hampered by the absence of access to information, capital sources, legal, technological and other related services. Services of this sort, applied effectively to energy-environmental technologies, would push technology cooperation forward at a dramatically accelerated rate. Such intermediary services could either come about through a new institution, funded with public and private capital, operating on the venture capital model, or they could come from expanding the functions of public agencies already responsible for promoting technology cooperation.

    Besides establishing an intermediary facility, some innovative strategies for increasing the available funds are needed, since the field of technology cooperation has been suffering from chronic shortage of financial resources. One strategy may be to use official funds such as the Global Environmental Fund as a leverage to make more effective use of private funds.

Concluding remarks

Our proposals are not meant to be exhaustive. A long list of measures needs to be assembled in order rapidly to promote technology cooperation which has not progressed as fast as expected.

Sufficient consideration has yet to be given to a number of basic issues related to technology cooperation. Such issues include the following matters:

- What is appropriate technology?

- To what extent and at what level should harmonization of standards be pursued (particularly concerning developing countries which have a great interest in this issue)?

- What kind of regional cooperation mechanisms should be worked out in order to address effectively regional environmental issues and the Long Range Transboundary Air Pollution issues?

- How could a 'global partnership' be created effectively in order to cope with global environmental issues?

- How can joint implementation – and, by extension, tradeable permit systems – become workable?

- How can financial resources be made available and more accessible?

- What new type of industrial development is needed in developing countries?

It is essential to establish a standing forum in which various groups, including governments, industry and NGOs, can regularly engage in discussion on issues such as those mentioned above. Preparations are being made for this type of forum through the cooperation of the World Resources Institute and GISPRI. We hope its establishment will result in the identification of opportunities and impediments in the field of energy-environmental technology cooperation and will deepen the understanding of the nature of the problem.

Appendix

Welcoming Address

*Michael Grubb, Head, Energy and Environmental Programme,
Royal Institute of International Affairs, London*

Appendix
Edited Welcoming Remarks to the Workshop

Michael Grubb

Let me start by thanking you all for taking time out of busy schedules to join us, and above all those of you who have worked to produce written papers and presentations. I want to take this opportunity to expand upon why we have organized this workshop.

Why are we here? Allow me to give a somewhat personal answer. In my research at Chatham House I have been most fortunate to have had the opportunity to interact with people concerned with energy and environmental issues from a very wide range of perspectives. That research led me to many conclusions. Among them I will point out just two simple ones that pertain to this workshop. First, the more I have studied energy from a global standpoint, the more I have become convinced that total physical energy resources are not the dominant constraint. Most of you will know that today, after more than a century of almost uninterrupted increase in global oil consumption, the proven remaining reserves are greater than they have ever been. The same is true of natural gas, only more so; and coal resources – and maybe economic reserves as well if we look globally – are greater still. To this I might add that I do not envisage that we will run out of viable uranium resources in the foreseeable future, and renewable energy resources vastly exceed our conceivable needs.

I do not deny that there are centrally important issues arising from the distribution of all these resources, which result in major and familiar constraints and complex policy issues. But I see the overall constraint not as total energy resources, but as the environment. By this I mean environment in the broadest sense of the quality of our surroundings, and the degree to which we can or will tolerate human interference with the natural systems that maintain local, regional and global ecosystems.

We can already see both physical and political stresses – and I emphasize both – at a time when global economic development is still at an early stage – I am tempted to say, at puberty. We will be considering some of the evidence for this more closely tomorrow, and I will leave you to dwell overnight on the analogy that global development is in the throes of difficult, confusing and, for many, upsetting changes which will culminate, I hope, in a greater willingness to bear responsibilities towards things beyond ourselves and our own creations.

I would only remind you of the enormous pent-up potential for increases in the scale of human activities and demand for energy and other resources, as four-fifths of the world's population aspire to the luxuries to which, within a few short generations, most of those in developed countries have become accustomed. Figure 1 gives an impression of the current inequalities in per-capita consumption of commercial energy, and the enormous potential for global growth as the developing world aspires to higher living standards, and

Figure 1 Energy consumption per capita and population, 1992

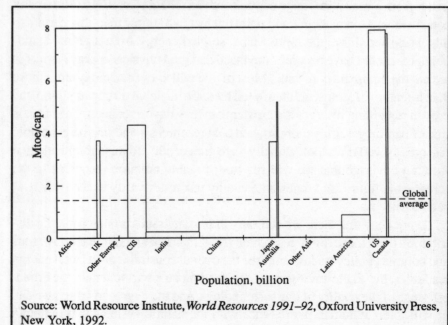

Source: World Resource Institute, *World Resources 1991–92*, Oxford University Press, New York, 1992.

as global population expands inexorably towards a doubling of today's level, which most experts consider to be the lowest level at which population stabilization can be predicted. It is a diagram worth absorbing in some depth, as it captures dimensions of both the total magnitude and the distributional characteristics of the global problem.

This is not to propose for a moment that development is a source only of problems. On the contrary, I would argue almost by definition that development is a source of both problems and solutions simultaneously. In this context, I believe it is important to eschew generalizations about the relationship between environmental degradation and development. The World Bank 1992 report on Development and the Environment (with a diagram reproduced as Figure 2) illustrated that very clearly. Some important kinds of environmental degradation – the most immediate and local – can decline from an early stage of economic development. Others, such as CFC emissions and atmospheric sulphur emissions, have initially grown with industrial development but have peaked and started to decline in richer countries. But we can identify at least three general environmental impacts which have escalated steadily with economic growth and are still doing so:

- The volume of material throughput and resulting generation of wastes (which include, of course, the wastes from other areas of environmental 'clean-up' like sulphur emissions)

- The impact on land use, including (but not only) associated loss of biodiversity

- Emissions of carbon dioxide.

Energy is involved in the first two, and is of course central to the third.

So much for the contention that environment is a more important constraint than overall resources. I would argue a corollary that extensive public policy on the environment is necessary and unavoidable, which leads me to my second contention, which will be less pleasing to the 'environmentalists' – if I may use that much abused term – here. This is that environmental policy has been developed with seriously inadequate regard for its impacts on industry or economy, and that this is becoming an increasingly damaging tendency which needs to be reversed.

Figure 2 Environmental indicators at different country income levels

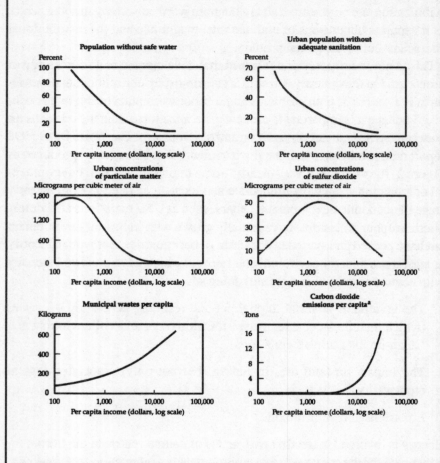

Note: Estimates are based on cross-country regression analysis of data from the 1980s.
a. Emissions are from fossil fuels.
Source: Shafik and Bandyopadhyay, background paper; World Bank data; reproduced from *World Development Report 1992*, World Bank/Oxford University Press, New York, 1992.

Environmental groups and policy have for too long existed in splendid isolation from other societal goals. We are here to consider the relationship with industry. Industry generates wealth and provides goods and services which generally, if not always, increase welfare – which is why people want to buy the goods and services. Industry needs profits to survive and to invest – not only in larger markets, but also in better and cleaner technologies and processes. Industry also benefits from stable and predictable policy, particularly as this affects its markets. Environmental policy has often been driven by the philosophy that it is the environmental goal which matters far more than the means, and often in appalling ignorance of the impacts on industry – and of how much the adverse impacts could be lessened by more coherent policy design.

It would not, of course, be fair to blame only one party for this. It takes two to tango, and industry has not always been exactly cooperative in the design of environmental policy. My central contention is simply that as the issues get tougher, the costs of bad policy become higher, and we have reached a point at which the quality of environmental policy matters as much as the quantity – perhaps more, because the strategic requirements of sustainable development cannot be achieved without cooperative, healthy and innovative industry.

I am conscious that these are not exactly new observations. But despite all the rhetoric along these lines, the process of dialogue towards improved policy is still not well advanced throughout the OECD, though I am also aware that some countries are further down the road than others. The central purpose of this workshop is to advance that dialogue and understanding. To take my earlier analogy perhaps further than I should, if both environmental groups and the energy industries have to pass through puberty towards sustainable development, we are here to consider what products might emerge from the subsequent liaison.

Other Publications from the
Energy and Environmental Programme

Books

* *The Earth Summit Agreements: A Guide and Assessment*, Michael Grubb et al, April 1993, £12.50 pbk, £25 hbk

Emerging Energy Technologies: Impacts and Policy Implications, Michael Grubb et al, June 1992, £29.50 hbk

The Environment in International Relations, Caroline Thomas, May 1992, £12.50 pbk, £25 hbk

Energy Policies and the Greenhouse Effect Volume One: Policy Appraisal, Michael Grubb, 1990, £12.50 pbk, £29.50 hbk. *Volume Two: Country Studies and Technical Options*, Michael Grubb et al, 1991 £12.50 pbk, £35.00 hbk

European Gas Markets: Challenge and Opportunity in the 1990s, Jonathan P. Stern, 1990, £25 hbk only

Reports and Occasional Papers

* *Power from Plants: The Global Implications of New Technologies for Electricity from Biomass*, Walt Patterson, April 1994, £9.95

The Struggle for Power in Europe: Competition and Regulation in the EC Electricity Industry, Francis McGowan, September 1993, £12.50

Evolution of Oil Markets: Trading Instruments and their Role in Oil Price Formation, Joe Roeber, October 1993, £12.50

Oil and Gas in the Former Soviet Union: the Changing Foreign Investment Agenda, Jonathan P. Stern, July 1993, £9.50

* *Environmental Profiles of European Business*, Dion Vaughan and Craig Mickle, April 1993, £15

Third Party Access in European Gas Industries: Regulation-driven or Market-led? Jonathan P. Stern, November 1992, £12.50

Paradise Deferred: Environmental Policymaking in Central and Eastern Europe, Duncan Fisher, June 1992, £10

Energy and Environmental Conflicts in East/Central Europe: the Case of Power Generation, Jeremy Russell, 1991, £10

Environmental Issues in Eastern Europe: Setting an Agenda, Jeremy Russell, 2nd edition 1991, £10

The Greenhouse Effect: Negotiating Targets, Michael Grubb, 2nd edition 1992, £10

The UK 'Coal Crisis': Origins and Resolutions, Mike Parker, October 1993, £7.50

Climate Change Policy in the European Community: Report of a Workshop, Pier Vellinga and Michael Grubb (eds), April 1993, £7.50

To order the above, and for further information about publications and the Programme's work, contact the Energy and Environmental Programme, Royal Institute of International Affairs, 10 St James's Square, London SW1Y 4LE. Tel: + 44 71-957-5711, Fax: + 44 71-957-5710. Publications marked * are also available from Earthscan Publications Ltd, 120 Pentonville Road, London N1. Tel + 44 71 278 0433, Fax: + 44 71 278 1142.